AUSTRALIAN PRECIOUS OPAL

AUSTRALIAN PRECIOUS OPAL

BY
ARCHIE KALOKERINOS

NELSON

THOMAS NELSON (AUSTRALIA) LTD
19-39 Jeffcott Street West Melbourne 3003
104 Bathurst Street Sydney 2000
and in Brisbane Qld, Adelaide SA and Perth WA

THOMAS NELSON AND SONS LTD
36 Park Street London W1Y 4DE

THOMAS NELSON AND SONS (CANADA) LTD
81 Curlew Drive Don Mills Ontario

First published 1971
Reprinted 1972, 1973, 1975, 1976

SBN 17 001960 8

Design by Barbara Tilzey

Typesetting by
The Universities Press Pty Ltd Melbourne

Printed in Hong Kong by
Leefung-Asco Printers Ltd

CONTENTS

7 Foreword

8 The Opal Fields of Australia

9 Lightning Ridge

14 Lightning Ridge Lives Again

18 Coober Pedy

27 The Nature of Opal and its Formation

30 The Colours in Precious Opal

32 Possible Colour Changes

34 The Basic Classification of Australian Precious Opal

36 A Study of Patterns

38 Solids, Doublets, Triplets and Chip Jewellery

40 A Miner's View of Opal

43 A Study of Faults

47 How to Value Opal

50 Index

ACKNOWLEDGEMENTS

The compiling of this book was aided by experts in widely different fields. Len Cram, a Lightning Ridge opal dealer, made possible an understanding of patterns; working with him led to the discovery of the 'searchlight' phenomenon. Bill Petrohelos, my ex-mining partner, showed how a disciplined approach could be made to the problems of evaluation. The same sort of discipline, forced upon me by Jack Oughton, a Sydney journalist, eventually led to the recognition of restrictions in colour changes, the types of colour in opal, the 'red light' phenomenon, and the classification of opal. Professor John Gunn, from the Department of English at the University of Sydney, through his interest in Australian terminology, was responsible for the realisation of this publication. The Physics Department at the University of Sydney, and the CSIRO in Melbourne and Adelaide provided scientific explanations for the phenomena observed. Harold Hodges, motel proprietor and opal collector, together with James Conomos of Walgett assisted with advice and the provision of specimens. To all the miners who provided specimens—often at considerable inconvenience to themselves—I owe a great deal.

The play of colour in precious opal

Common opal does not display a play of colour

Shrinkage of rock producing space for opal deposition

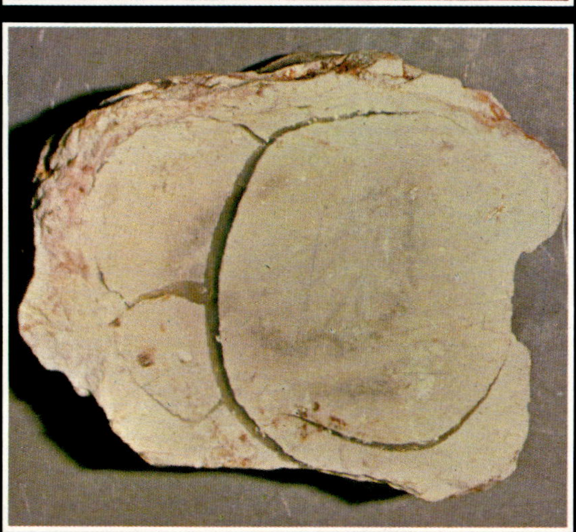

Opalised shell

Opalised bone—solid precious opal

Opalised wood—the precious form

Water-washed, rounded pebble stone in opal

Sharp-edged, rough shaped sandstone chunks in opal

Horizontal banding in opal pattern formation

Bending of horizontal banding

Drip formation of patterns

Concentric formation of patterns

Horizontal banding, concentric pattern formation and drip formation—in one specimen

Colourless transparent potch

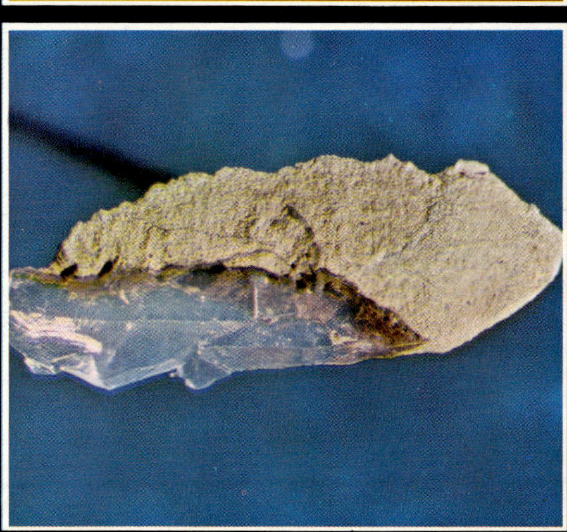

White potch

Grey potch

Black potch

Amber or 'beer-bottle' potch

Blue-bottle potch

Potch-and-colour

Spectrum colours from violet to red

The searchlight phenomenon. The opal is hand held

Spectrum red changing to spectrum blue

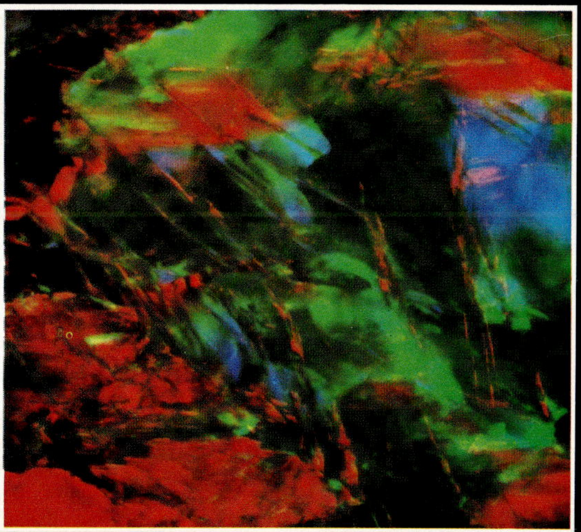

Spectrum colours in precious opal

Pastel shades of spectrum colours

'Marine' shades of spectrum colours

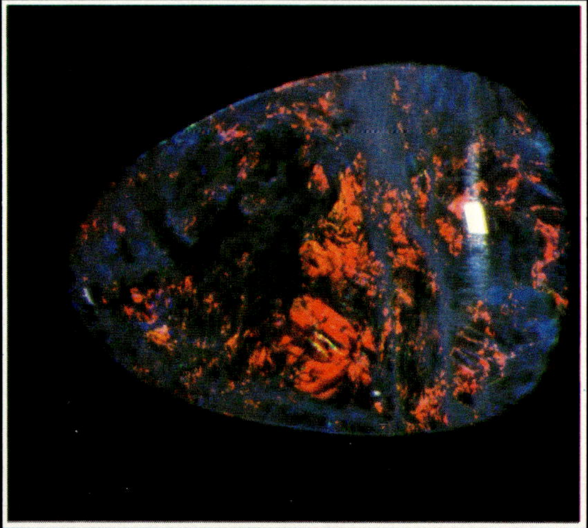

Red on blue

Red on black

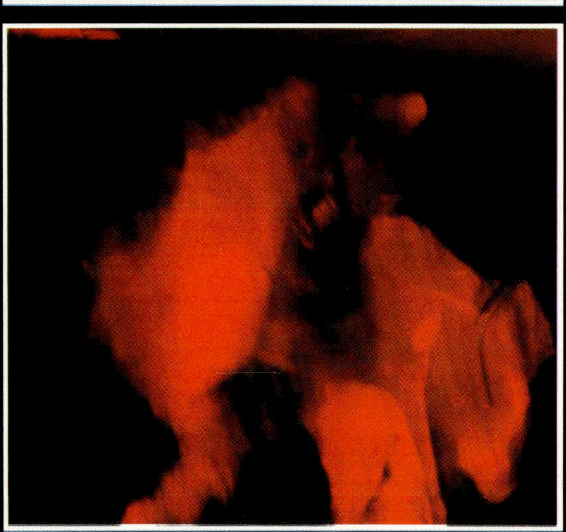

Red coming out of the black

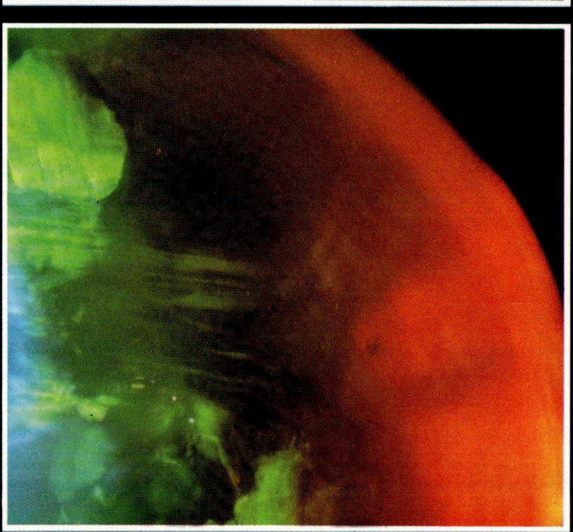

The red light phenomenon, same specimen as previous two

The red light phenomenon showing on surface irregularities

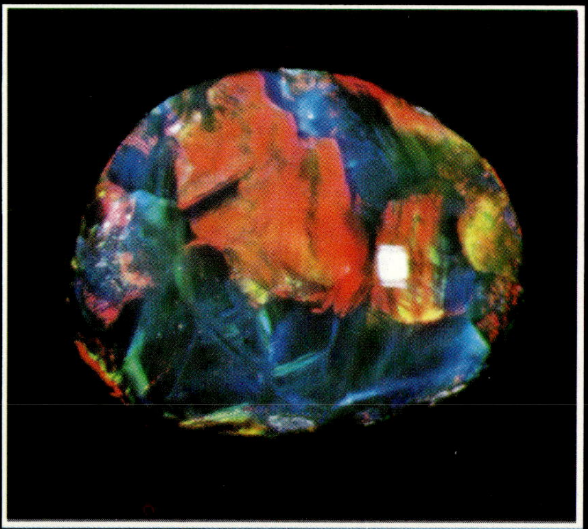

Crystal black opal

FOREWORD

The magnificent splendour of the Australian opal that has fascinated the world for almost a century is still clad in a mysterious veil of fear and superstition whose only explanation is the lack of knowledge.

Much has been done in recent years to lift this veil, but the deep understanding displayed by the first pioneers of the opal fields has not been sufficiently appreciated.

These men loved their country—the deserts of the inland and the parched red ranges where Aboriginal tribes fought to preserve what had been theirs since the beginnings of dreamtime. They loved the open spaces and the nearness of the stars, the scent of gidgee burning and the freedom that was theirs. But most of all they loved the opal, in a tender expert way that enabled them to learn so much about it and impart their knowledge to men from opal fields scattered throughout Australia.

That is how the terminology of the opal originated. It is a beautiful terminology, full of meaning to those who 'know' but quite incomprehensible to the scientific minds of men who sit in judgment in the comfort of the cities. It functions amazingly well in practice, provided that the original concepts are kept in mind. There is a mass of knowledge conveyed by a few simple words. For example, to speak of 'crystal' opal is meaningful, but to speak of 'crystal boulder' opal is so much more meaningful.

A book on opals, like a book on medicine, cannot be complete or entirely accurate. To become an authority one would need to spend a lifetime working on each and every field—obviously an impossibility. I do not hesitate to admit deficiencies in this book—some of which may be serious—but to delay the publication until these are corrected would be to delay forever.

THE OPAL FIELDS OF AUSTRALIA

Scattered throughout the inland of Australia are the opal fields. In actual fact, much of the inland is opal bearing, but only parts are rich enough to warrant serious mining. Some opal is found in mountainous and coastal regions, but this is of academic interest only.

The main fields of Coober Pedy and Lightning Ridge are separated by 1,500 miles. They are as different as chalk is to cheese. The country is different, the men who work them are different, and the opal is different.

Lightning Ridge in New South Wales produces most of the black opal seen in the world today. It also produces considerable amounts of light coloured opal. Coober Pedy, in the far north of South Australia (sometimes called 'Central Australia') is famous for its big 'runs' (patches) of light opal. Although not valued as high as good black opal, the quantity found is large indeed, therefore there is more money there.

White Cliffs in New South Wales is virtually worked out. The boom days there were at the turn of the century. The opal found was good quality light material.

The Queensland fields remain mysterious. Seventy years ago many of them boomed. Drought and hard times closed most of them down. There is no doubt that some will boom again, but economical prospecting has not been possible. The draw card is quality and reliability so that miners and dealers watch any development with keen eyes.

Mainly because my experiences were there, I will discuss in this book Lightning Ridge and Coober Pedy only. This does not mean that other fields are not meritorious; it only means that Australia is a big, big, country and the individual opal fields so detailed and absorbing that one would need to be as old as Methuselah to even sample the truth that is in them.

I have often said that I could go to any opal field in Australia and come back after three months with a story that would be different to anything yet described; I hope to do this in the following pages.

LIGHTNING RIDGE

There is only one Lightning Ridge and it is doubtful if an area even remotely similar will ever be found. The very nature of opal formation means that worthwhile fields exhibited surface indications (that is how they were found). Since most of the earth's surface has been prospected the continuing supply of opal over a long period is very doubtful indeed. This means that gemstones being mined at Lightning Ridge today will become priceless heirlooms.

Black opal does exist elsewhere but only at Lightning Ridge does it reveal itself with so many unique and magnificent features. There is no end to its variety; each stone is an individual; its twin does not exist. Its entrancing beauty makes men fight to find it, to possess it, to keep it and fondle it with care. When somebody else unearths a perfect stone, jealousy is inevitable.

Opal and talk of opal permeates the air at the Ridge. Occasionally, when big strikes are made, the atmosphere becomes electric. The flashing reds on purple-blackness, the golds and orange-greens, the rolls and squares of fire light up the world making all else seem so drearily unimportant.

To be a miner at Lightning Ridge qualifies one for membership of a very special club—the club of élite opal lovers. Even its jargon is something remarkable in itself. Strange words are used to describe the opal; sometimes there is no need for words—one look is enough to the eyes of a man who shares the secrets of the unsurpassed beauty of the stones. Exaggerated? Certainly not! But a word of warning, because this magic cannot be grasped instantly. Like so many of nature's gifts, it must be earned, and that takes time.

The town of Lightning Ridge is booming by Australian standards. Its population of about 1,500 consists of miners, station hands, shearers, fencers, ringbarkers and men like store-keepers who provide essential services. There are modern airconditioned motels, a bowling club, progressive associations, a bush nurse, a new school, and some full-time opal cutters and dealers. It is a wonderfully democratic society because all men, successful or not, dress, think and talk alike. There are unwritten rules that decide man's position in the hierarchy of this society.

Respect goes first to the few remaining old miners, for these men know the opal. Next in line are the men who find 'named' stones. Such stones are of sufficiently high quality to warrant special names. In my own time I have seen 'The Cardinal', 'The Red Robin', 'Preston's Stone' (named after the finder), and many others. Ten years ago I could have purchased some of these for a few paltry pounds. Today the mortgage on a mansion would not be sufficient. Each time a named stone is found it seems that the ultimate has been reached, that nothing more perfect could ever be found. And so it is, yet nature has a subtle way because the next named stone will be different—neither more perfect or less perfect, just different.

One wonders about stones found years ago, like 'The Pandora', 'The Butterfly', 'The Pride of Lightning Ridge', and 'The Light of the World'. They must have been truly magnificent. Those who love the opal and the land that gave it birth are hurt to see such gems disappear to foreign lands forever.

There is no need to be nostalgic about the past as far as opal is concerned because a miner working at this very moment could easily have in his hands tonight another named stone, and it could come from a patch of ground previously ignored. Why is this so? What made the opal at Lightning Ridge and what indications are there to guide the miner? To understand this, we must go back to prehistoric times, when nature deposited the rocks in which the opal now rests.

Geology, at Lightning Ridge, from a miner's point of view

Scientists cannot agree as to when the various layers of rock at Lightning Ridge were deposited. There is no doubt that this was a relatively recent event and that the rocks are sedimentary.

An inland sea, or lake, fed by rivers deposited the layers up to a total depth of about 100 feet. Since these layers contain the silica that made the opal the first restriction of opal formation becomes apparent: opal is limited in extent from the surface to a depth of 100 feet.

What followed the deposition of the rocks is mostly unknown. All so-called 'rules' discovered by miners have eventually been disproved. What applies in one area may not apply in another only a few short feet away. However, certain features are relatively constant.

The first of these can be seen in many places. It is regarded as typical of opal-bearing country. In fact, miners seek it first because without it the chances of finding opal are slim indeed. In such places a clay-like layer (known as 'opal dirt') was deposited up to a depth of a few feet or more. In its present condition this is usually damp, soft, and easily dug with a pick. The top few inches or more is usually darker in colour (brown and purple) and is known as the 'level'. Sometimes the level may be dry and very hard.

Above the level is a harder rock known locally as 'sandstone'. Its thickness varies from a few feet to many feet. In places, separating the true sandstone from the level, is a hard layer known as the 'steel band'. This varies in thickness from minute to several inches.

If opal is present, it is usually in the level, but it can be anywhere in the opal dirt too. The presence of a steel band is regarded as a favourable sign, particularly if it 'makes' and 'breaks', that is, if it forms, breaks up after running a short distance, then re-forms again.

In some areas the pattern of opal dirt and levels covered by sandstone was repeated, like a sandwich, up to eleven times. In a particular area, therefore, there may be one or several levels. Here and there the sandstone reaches the surface, but usually it is covered by other types of rock such as what is known locally as 'shincracker'. Mostly, the first few feet from the surface consist of irregular conglomerates.

Unfortunately, the picture is not always as simple as the one just described. The first variation is the occasional finding of levels not covered by sandstone roofs. They may have been formed in this way but it is usually assumed that originally sandstone did exist and was later washed away. The levels in such cases appear on the surface or are covered by secondary sediments.

Next, opal may be found mixed with gravel-like surface layers. This is always at the foot of hills containing opal and it is assumed that opal levels on the hill were eroded, permitting the opal to be washed down into the surface gravel.

Now we come to the big problem of opal formation at Lightning Ridge. Levels are common, but opal is not. What then, are the other factors involved? This question cannot be fully answered, but we have certain relatively consistent indications. It is believed that the sandstone provided the silica that was dissolved by water, forming a gel; ground movements resulted in space formation, permitting collections of silica gel that later solidified as opal.

Signs of ground movement can be seen and these, in most areas, bear a definite relationship to opal formation. There are 'slides', which are really faults where the ground cracked as it moved. Levels often rise or fall in the vicinity of slides (because of ground movements) and the rocks on one side may be different from those on the other (for similar reasons). Slides may be almost vertical or almost horizontal. There may be one slide in an area or several scattered in an irregular fashion and varying in nature as far as relative ground movements are concerned (that is, the direction of ground movement). A slide may be only a few feet or several hundred feet long (a 'king' slide). The area extending away from where a slide faces upwards is known as the 'front' of the slide. The opposite area is referred to as being 'behind' the slide.

While no definite rule applies, it can be said that if opal is present it is either in front of or behind the slide (there is considerable argument about actual distances). In other words, no slide, no opal.

The presence of multiple slides appears to influence the formation of opal, but no definite pattern has been observed. One slide may make opal by itself but elsewhere two or more slides are involved. Here it must be stressed that only a small proportion of slides make opal and only a small proportion of the opal is precious; the rest is potch.

In some areas the sandstone roof drops suddenly at a right angle for a few feet or more, then continues horizontally. Known as a 'wall', such a drop is regarded as a favourable indication. Opal may be found on its high or low side. Walls may be anything from a few feet to several hundred feet long.

The next indication of opal formation is known as a 'dome'. The sandstone roof may be shaped like the inside of a dome. Opal tends to occur in the levels beneath this.

Confusion surrounds what are known as 'blows'. Some miners use the term when referring to any form of sandstone appearing on the surface. Others restrict it to where it appears that chunks of sandstone were 'blown' to the surface through the deeper layers. Below ground evidence in support of this can be seen in the nature of sandstone chunks mixed with opal dirt and other rocks and penetrating all layers. It is known that opal may be formed in the vicinity of such blows.

Breaking through to the surface in some areas is what is known as a 'biscuit band'. The name was suggested by the flakey, biscuit-like layers. Most miners consider that they consist of slides of opal dirt. They may be five inches or more in thickness, appear to go down into levels of opal dirt and may carry good opal. In some places they have been found penetrating into black soil.

The history of Lightning Ridge

After its formation, the opal lay resting beneath the sandstone beds for uncounted ages. Aborigines sometimes stopped on the ridge while hunting kangaroos to sharpen spears or gaze across the north-west plains. They must have noticed coloured flashes where levels eroded to the surface and they must have found some opal stones. What they thought of them we will never know, because nobody can remember and those days have gone forever.

One hundred years ago white men came, bringing flocks of sheep and the sound of ringing axes to the district. Originally, they called the area 'Wallangulla', but a drover noticed how lightning was attracted to the low red ridge and renamed it 'Lightning Ridge'. A more fitting name for the birthplace of stones of fire would be difficult to imagine.

In 1902, amongst the shearers, station hands and boundary riders working near Lightning Ridge, were two men whose destiny became history. The first of them, Fred Bodell, is still at Lightning Ridge. With a clear memory he can tell about the second man, Charlie Nettleton, who is credited with the discovery of black opal.

Before Charlie Nettleton came on the scene it seems certain that opals were found on the surface by Aborigines, wives of settlers, and others, but these folk did not realise the significance of their findings. Charlie Nettleton, however, knew something about opal. Stimulated by what had been found he sunk his first shaft about 1904; others followed, mainly from the former opal fields at White Cliffs. By 1907 Lightning Ridge was declared an official mining area.

One would expect the black opal to have achieved instant recognition, but that was not so. Puzzled buyers shunned it. Most regarded it as a form of 'matrix', 'unfinished' opal, or something so queer that it was best avoided. This unbelievable situation, where the world's finest gemstone was thrust into ignominy, persists to some extent until the present day. Worse still, the inheritance of terminology and understanding handed down to us by the pioneers of opal has never been fully appreciated. A wonderful new form of art—nature's last and finest effort—had been opened at last for all to see, but the world did not appreciate it.

Stones sold of course, at ridiculously cheap prices, with natural domes of unseen beauty ground to horrid flatness, with shapes and patterns destroyed forever.

Fortunately, the beauty of the opal affected many of the early miners. Bewildered by the attitudes displayed by outsiders they retreated into a little world of their own. In tents and bush huts made of branches, over campfires where the scent of gum leaves mingled softly with the darkness, they talked and thought and dreamed. With tenderness and love, with an understanding that seems incredible for men who lived so roughly in the semi-desert outback, they learned about the opal.

Mining in those days must have been difficult. Picks, hammers, chisels, leather or iron buckets, and a hand windlass were the only tools; light was provided by candles stuck into shaped pieces of wire that had been coiled at one end and, skewer-like at the other, were pressed into the mine wall. Known as 'spiders', these remained a feature of Lightning Ridge life for over fifty years.

As a shaft went down, the removed dirt collected on the surface. Simple means of transporting away the dirt did not exist; timber log cabin-style retaining walls were built around the shaft entrance in order to prevent dirt falling back. Thus, the height of the shaft opening above the surface of the ground increased as the mine deepened.

If opal dirt was found and the 'country' looked promising, horizontal drives would be made beneath the sandstone roof. The dimensions of these were severely restricted, because removing dirt was tedious. Miners would crawl into burrow-like drives, gouging slowly in cramped conditions, because one false move with a pick could destroy the only opal of value in the whole claim.

Rich claims were 'emptied' out, that is, as much of the opal bearing level as possible was removed. When wide drives were necessary, 'pillars' of untouched dirt were left as roof props. As miners gouged, the opal dirt was always at their feet and became known as the 'toe' dirt in contrast to the opal bearing level known as the 'pay' dirt. In some mines it was possible to stow some of the dirt removed in disused drives—hence the term 'stowed dirt'. At the time, small stones were not profitable; combined with the inefficient mining methods, this meant that they were either neglected or missed and dumped with the surface or stowed dirt. In recent years, these small stones have become popular and miners have sifted surface and stowed dirt to recover them; pillars have suffered a similar fate.

The characteristic ping as a pick hit a nobby was the miner's guide. Nobbies were scattered in the levels like currents in a cake, only much more thinly. Sometimes, collections known as 'pockets' would be found. Under these conditions the pattern of a miner's work would alter: he would be careful and slow, thus informing indirectly other miners that he had struck it rich. If confirmation was needed, other miners could examine the dirt hauled to the surface and find in it 'traces' (chips or thin pieces of precious opal).

This alerted the 'ratters', the scavengers of the opal fields, who watched like hawks until rich mines were left unguarded. To nimbly skim down a shaft, smash a few opals from the level and disappear into darkness, was easy for these most hated of all law-breakers.

Court records, now in the possession of Harold Hodges and kindly lent to me, provide an insight to those times. There is no mention of ratters being caught, although it is common knowledge that they existed. It is possible that miners held their own 'kangaroo' courts, passing sentences of banishment, if necessary complete with tar and feathers.

Conflicts between squatters and miners were surprisingly few. Water supplies provided the basis for some of these. Holes, called 'tanks', were sunk into the ground at low places to collect rain water for stock. Since this was all that was available, miners were forced to 'steal' some of it for their own use. Eventually the government sank a special tank for miners and peace was restored. Of course, some miners also helped themselves to a few 'stray' sheep, fresh meat being regarded as superior to bully beef or no meat at all.

According to the records, the now well known author, Ion Idriess, obtained a miner's right at Lightning Ridge in 1909. Fred Bodell is mentioned too, but the reason for his fame is not. Some time during those early days he spun a halfpenny into the air and advised a friend to dig where it came to rest. Beneath that spot the Pandora lay waiting—590 carats of unmatched stateliness.

The next period in the history of Lightning Ridge was not a progressive one—in fact there was a sharp decline. To understand the reason of this decline requires a knowledge of certain geographical and geological features. The importance of this knowledge, however, goes beyond the past. Intelligent and business-like prospecting today depends on it.

Rising approximately sixty feet above the black soil plains, the Ridge itself is several miles wide and really unlimited in length, because it is part of a ridge system that extends through a vast area of New South Wales and Queensland. It so happens that opal was deposited mainly in patches. Known as 'fields', these varied to a very high degree in extent and richness. Some, like the Three Mile, are more than a mile in diameter, while others are only a few yards. In big fields the opal is not continuous: it makes and breaks and, as luck would have it, the breaks are in the majority.

In most places rich fields exhibited surface indications and that is how they were found. Miners flocked by, finding large amounts of good quality opal. Shafts were sunk by the hundreds in closely packed patterns. Then, as the opal thinned out on the edges of the fields, shafts were widely spaced or not sunk at all. Thus the area around Lightning Ridge became distinguished by fields pock-marked with shafts and gaps of varying extent until neighbouring fields were reached.

The time came in the early 1920s when all the rich and easily worked fields were exhausted. To prospect new fields was not economical, because without surface indications there were few guides and, anyway, on such places opal was not thickly deposited. The Ridge slipped backwards, most miners left the field, while those who remained barely scraped a living. Some supplemented incomes by working part-time as shearers or station hands. Older men became pensioners living off government allowances. In fact the Ridge consisted mainly of pensioners.

Life in the area was quiet at the time. In the pub and under the Tree of Knowledge (an old tree at the junction of two streets at the pub corner) the inhabitants droned the years away talking about the 'good old days', yet enjoying this outback township where the restrictions of civilisation seemed far away.

Things were very much the same when I came to the area in 1957. I remember how astonished I was to find the pub open (by law it should have been closed). Furthermore, Aborigines were in the bar drinking with Europeans (this was also illegal at the time). 'We make our own laws here', somebody told me, 'and we treat all men as equals.'

Most of the miners lived out on the fields, in huts, shanties, and tents. There were dogs and horses, a few old cars, some trucks and bicycles. Many miners lacked even the simplest means of transportation and—they walked. The bleached white heaps of sandstone looked like monuments to better days, gone forever. The few opals flashed from dirty tobacco tins beneath my eyes failed to impress. It seemed that nothing could revive the Ridge; the place was a museum.

How wrong I was!

LIGHTNING RIDGE LIVES AGAIN

In the early 1960s there was a sharp increase in the production of good quality light opal from the South Australian opal fields of Andamooka and Coober Pedy, resulting in the establishment of a vast market and dealer network. At the same time 'puddlers' were introduced to Lightning Ridge. These were mechanical sifting machines designed to sift opal dirt, separating nobbies and making possible the economical handling of large volumes. They were ideally suited for sifting surface and stowed dirt and this was done on a rapidly increasing scale, the opals found being quickly absorbed into the market network established for light opal. On top of all this, the Ridge was caught up in the natural progress of rural Australia. Water and electricity were installed, roads improved, and the rigours of outback life alleviated considerably.

Puddling became fashionable. Old cars, tractors, and machines of all sorts were converted into puddlers, some large, some small, with fine white dust flying everywhere. Astute 'miners' (they were now really 'puddlers') inquired as to the whereabouts of mines that were rich in the old days because around them the surface dirt would be rich and the stowed dirt very similar. Opinions of old timers became much sought after, the 'fee' being a few schooners of beer.

An odd, proud miner still worked in the old way, gouging into the few small areas of untouched ground on old fields or prospecting in relatively virgin country. Failure invariably followed true prospecting: those who tried it went broke and had to return to puddling or working on old fields. The problem was one of shifting dirt from below ground to the surface. This could not be done quickly enough or economically enough to warrant true prospecting. Everyone wondered what would happen once all the old dirt was puddled. The Ridge had boomed a little, but the long term outlook was far from bright. Fortunately, the introduction of two new machines altered everything.

First, streams of water were used to assist the process of puddling. This meant that miners did not have to wait for opal dirt to dry out, puddling was quicker, more effective, and less damaging to fragile stones. The building of a special miner's puddling dam, fed by bore and rain water, partially solved the problem of water supplies for these machines, but water is still a little scarce. Puddling machines that used water became known as 'wet' puddlers, those that did not, 'dry' puddlers.

An even more important innovation was the automatic self-tipping hoist—a cheap effective unit that lifted buckets of dirt from below ground to the surface, automatically emptied them and returned them below ground. One man could work alone with the aid of such machines. Usually, the toe dirt was loaded onto a truck and disposed of on waste ground while the pay dirt was puddled. Gouging was unnecessary; jackpicks could remove the pay dirt in bulk; a whole claim could be emptied in a matter of weeks; modern mining had come to Lightning Ridge.

Of course there were those who could not afford even simple machines. They used their heads, studied the country carefully, listened to the talk of others and waited, worked, and waited . . . With hand tools only, their shafts went slowly down to bottom as one man did—on the Red Robin.

In a way I object to the publicity given to this priceless gem of fire, for I held it in my hands one day and would like to think that it is mine, not seen or felt by others. It is red, of course, though that is not how it received its name: a bird, a real red robin, was there when the miner pegged his claim. I like the name; I love the stone; I know that more are there. Some, no doubt, will outshine the Robin in details of perfection, but I will always remember how blue the sky was and how the eucalypts sang on the day I saw the Robin.

The fields at Lightning Ridge

Opal can be found in the main street of Lightning Ridge if one cares to dig (which the law forbids). Otherwise one can go on to one of the fields that spread away from the town on every side. Some fields, like the Three Mile, the Four Mile, the Six Mile and the Nine Mile, are named according to the respective distances from town (though not necessarily in the same direction). Others, like Canfell's, Walshe's, Paddy Burk's and Holden's, are named after those credited with their discovery or the owners of the richest claims. A few, like Thorley's Six Mile and Rouse's Six Mile combine both features. Then there are fields like Hornet's Rush, Bishop's Rush, Mulga Rush (mulga is the name of an inland tree), Western Fall (where the Ridge falls westwards), The Gully, Hawk's Nest, Old Town, New Town, Old Chum, New Chum, Potch Point (often humorously called Potts Point after a fashionable Sydney suburb), Poverty Point, Nobby's, Canada, Frog Hollow, Frying Pan and many others.

Rich old fields are distinguished by large numbers of shafts surrounded by high dirt dumps. Most of these dumps were originally even higher. In recent years puddling and 'noodling' (scratching in the dirt in the hope of finding opal) have reduced their heights.

New rushes are distinguished by an air of activity, though this may not, at first, be obvious. Dozens of claims will be pegged around the mine that started the rush. In recent years new fields have not contained large numbers of rich claims, but this is overcome to some extent by the quickness of modern mining methods.

So you want to find opal!

In New South Wales the possession of a miner's right (costing one dollar for six months or two dollars for one year and obtainable at police stations or mines department offices) gives the owner the right to prospect, peg a claim not more than 100 feet square, and work that claim. There are certain restrictions of course: one cannot, for example, dig on the front lawn of Government House or in Pitt Street, Sydney, yet otherwise the law is very much in the miner's favour. Once a claim is pegged, it can be registered to prevent other miners from taking it over. There is an initial cost of $1.40 to cover this but it is perpetual as far as an individual miner is concerned and it applies to all future claims. There are laws that cover the mode of pegging and working claims. 'Working' means the involvement of one competent person per claim (to stop big company take-overs). At Lightning Ridge there are rarely disputes concerning working laws as far as individual miners are concerned. Frail old ladies and men 'own' claims and actually do very little work but nobody bothers them. When big companies become involved, the picture may be different.

In some areas it is difficult to determine if a particular piece of ground is already pegged as an official claim. There may be pegs and notices, but these may be meaningless, the miners involved hoping that nobody will bother to question legalities. There is a move afoot to survey the area properly and this will allow proper records that can be checked.

If one does not wish to bother with the formality of pegging a claim then work is still possible, but no legal protection exists as far as attempts at taking over are concerned and anyone else can enter and work on the area at the same time. There is, however, plenty of unclaimed ground. Visitors who want to try their luck can noodle, dig, sink shafts, puddle, or do as they wish according to the available equipment (much of this can be hired).

Serious mining is also carried out according to available equipment. This need not be elaborate but naturally, the better the equipment, the more efficient the work. Ordinary picks and shovels can sink shafts and are used by many. Others enlist the aid of gelignite, air compressors, jack-picks and electric drills. One company introduced a giant mechanical drill that sinks wide shafts up to ninety feet in a few hours. When a shaft becomes too deep to permit the throwing out of the removed dirt, some other method must be employed. A hand windlass will do. Full buckets can be 'fished' out if one is working alone or wound up by a partner. Mechanical hoists are, of course, more efficient. The automatic, self-emptying type have already been mentioned.

Driving beneath the sandstone roof can be done with ordinary picks and shovels, electric jack-hammers, air operated jack-hammers and gelignite. The dirt is usually loaded into buckets, about twelve gallons in capacity, and carted to the shaft bottom in special wheel-barrows. So far, more elaborate methods of driving underground have not been used at Lightning Ridge.

Usually, the toe dirt is removed first. Then, according to equipment available, the level is either carefully gouged underground or taken to the surface in bulk and puddled.

Some miners use 'rumblers'—horizontally arranged machines that turn and permit the bulk of loose dirt to fall through leaving only small volumes that are puddled. One advantage of this method is the saving of precious water.

Shallow ground can be worked from the surface. So-called 'alluvial' opal on fields like the Nine Mile Flat where the opal has been washed by rain into the surface gravel can be found by careful digging or puddling. Trench digging has been tried. Bulldozers are used to a limited extent.

But no matter how much work is done, how efficient machinery is, or how much knowledge an individual miner has, the most important element in finding opal is luck. One can never really work enough or buy enough machinery to reduce luck to a size of no importance. Time after time astute miners have been 'certain' that opal would be found. They have dug through 'beautiful' sandstone (just what is good and bad sandstone seems difficult to determine) into levels 'fit to eat' and no opal has been produced. On the other hand, a new miner has often astounded experts by finding opal in levels that are too 'hard', too 'soft', too 'dry', or too 'wet' to warrant more than passing attention.

Opal is where you find it, as one man learned only yesterday. While I was typing the manuscript of this book, he unearthed a ten and a half carat stone. The rumour is that this is 'it', the 'best', the finest of them all. By now the news will be half-way round the world in offices as far apart as Tokyo and New York. A few overseas dealers will even think of flying all the way to Australia to see this stone for themselves. That is what black opal can do and that is why the Ridge will live forever in our history.

Selling black opal

Lightning Ridge opal is rarely sold 'in the rough' (before cutting) for the simple reason that it is impossible to forecast the appearance of finished stones. Nobbies may (in fact, one could say, almost always) contain sandshot faults, the colour may not 'come up' as expected, it may not 'look at you' (only appear from certain critical angles), or brilliance may be lacking. These faults seriously lower values. On the other hand, if a miner sells an uncut stone cheaply it may cut into a priceless gem.

It is the miner's prerogative to state his price. He will usually obtain this with the aid of the cutter's experience (most cutters are paid a percentage of estimated value). He may ask the advice of friends or value a stone himself. If the dealer thinks that the price asked is correct he will buy. Much depends on a dealer's outlets. One particular dealer may be able to dispose of certain stones that another may not. Prices are calculated in pounds (miners still calculate mentally in pounds) or dollars per carat—ranging from less than 1 to over 1,000.

Most miners add about twenty-five per cent to calculated values before stating prices. If a dealer's estimate is near the figure asked some bargaining may take place. If the price asked is far too high for an individual dealer he will usually simply state that he cannot buy, and the matter comes to an end. Rarely in such circumstances will an experienced dealer offer his opinion as to value.

Sometimes the beauty of an opal blinds a dealer to

imperfections. If he is experienced he will avoid this by being quite 'clinical' or impersonal. He is there not as an opal lover (though the chances are that he loves every stone) but as a man of business. If he can see a profit in a stone he will buy, otherwise he will not.

About seventy-five per cent of stones are sold on the field at Lightning Ridge. Most of the remainder are sold to dealers in Sydney and Melbourne. At the time of writing there are about twenty-one regular local and visiting buyers of Lightning Ridge opal.

A stone that is sold by a miner for $100 eventually sells to an overseas or local customer for $400. In the meantime it may pass through the hands of one or more other dealers, overseas agents, manufacturing jewellers, wholesalers and retail stores.

Unofficial estimates suggest that the Ridge derives an income from opal in the vicinity of $2 million per year. Allowing for dealers' profits permits an estimated $3 million added to Australia's income per year. There are 400 miners registered in the association at Lightning Ridge. In winter there may be between 600 and 800 miners working full or part-time. The average income of a miner, therefore, is not very great. Fifty per cent appear to 'do well'. Thirty per cent manage to scrape a living. Twenty per cent walk away from the Ridge penniless.

Importing opal into the USA is simple. One pays a four per cent import duty and the opal can be sold anywhere to anybody. In Japan opal can only be sold to the holder of an import licence and opal cannot be freely taken around the country. It may take up to one month to clear the material through customs.

COOBER PEDY

Heat, dust and flies; miles of sunburnt plain; saltbush and gibbers; a cloudless sky; a pattern of mine shafts surrounded by white sandstone dumps; a few crazy tin sheds and dugouts that serve as houses; and a never-ceasing wind. This is Coober Pedy—for some a place of treasure, for others a living hell.

Most of it is on a plateau—a sixty to eighty feet rise that spreads to the east and never seems to end there because the fall is gentle. Westward, the drop is sudden and this, meagre though it is, creates the inspiring beauty of the Stuart Range.

It is all desert. Sometimes, when rain comes sweeping from the north, water-holes fill; wild flowers bloom on a carpet of spotted green but this is soon forgotten when the water has gone and nothing but dust remains with the gibbers that never change.

Men died before they found this place. They came from the south on pack-horses and camels in search of a sea but they did this in vain because there is no sea, no creek, no spring, not even the shade of a single tree—only the desert rolling on and on.

To describe this country is difficult; to photograph it is impossible. Only an artist can reasonably show what is there, and the colours he needs are few—brown for the rock, red for the dust, grey for the bush, blue for the sky and white when the clouds sometimes soar and the sandstone glares in the shimmering heat.

One can only love or hate a country like this; to remain neutral is impossible. On some it casts a spell, captivating them for life in a network of loneliness broken only by the necessity to maintain some contact with others, to pay income tax, to register motor vehicles or buy petrol, food and stores. These people are happy only in the desert. They shun the squalid confines of the cities and the burden of too many people in too tight a space. Their heaven is all around them—in the distant reaches of the plains and the glory of the stars.

Others have a dread, an inborn fear and hatred of places so flat, so vast and so far from the sea. The loneliness awes them, the silence depresses. They cherish instead the bustle of the traffic, the dazzle of a thousand man-made lights and the parched, gaunt faces of a million strangers who somehow comfort them.

But whether one loves or hates has little bearing on why so many people go to Coober Pedy. The lure is not the inland or escape from the maddening throng. It is the ever hopeful fact that each time a shaft is sunk up to ninety feet into the sandstone beds there is one chance in a thousand that it will bottom on a 'run' of opal worth $200,000 or ten times more. At Lightning Ridge it is the beauty of the opal that controls most of everything. At Coober Pedy it is money and the men who come to get it.

Created thus is a fantastic air of excitement, adventure, thrills, romance and, of course, trouble. Nothing ordinary seems to happen in Coober Pedy, for even when it begins that way something soon changes it.

Visitors rarely get a glimpse of this way of life. Even journalists fail to penetrate it as books, magazines and newspaper articles suggest. I have seen some of the most adventurous men in Coober Pedy whose ordinary day-to-day lives would thrill a schoolboy and whose more turbulent times would be unprintable, photographed by journalists and simply labelled 'a miner' or 'an Aborigine'.

Central Australia still belongs to the Aborigines in a religious and tribal sense based on the sacredness of land through legends, the tenderness of nature, manliness and goodness in spite of some apparent cruelties. Occasionally, when tribal death sentences are passed or men die mysteriously in the desert, opal miners are awakened a little to the realities of Aboriginal life. At other times prospectors will try to 'con' Aborigines into revealing the whereabouts of rich opal-bearing land, or they will try to borrow the hawk-like eyes of Aborigines to spot odd but extremely valuable opal stones that are mixed with surface dirt in many parts. Otherwise the Aborigines are ignored—regrettably, because they can teach us so much.

Geology, applied to mining at Coober Pedy

The story of Coober Pedy is intimately tied with geology. Some of the most fantastic episodes of human action and reaction that I have witnessed have come about simply because of the way that nature deposited opal in Coober Pedy. Furthermore, the great mystery of opal deposition and formation is illustrated in its purest form in Coober Pedy. It is there for all to see. Illiterate miners from all parts of the globe recognised it, lived by it, dreamed about it, fought over it, and gloried in it, but technical authorities, geologists and scientists, let it pass unnoticed. How strange this is, how incredible but true—and to think of all the money that *might* have been mine if I had this knowledge six years ago or applied it later with more intelligence.

As in Lightning Ridge, rocks were deposited as sediment in the beds of lakes or seas. In some places a layer of special rock known locally as 'sandstone' was deposited. On top of this a layer of darker and muddy looking material was deposited for a thickness of a few inches or more. This became the opal bearing 'level'. In turn it was covered by another layer of sandstone which was covered by another level and so forth up to six or more levels over a total depth from the surface of about a hundred feet. Space for opal deposition was provided in a very special way.

The various layers just described were laid down more or less horizontally. Then the final movements of the earth's crust occurred and it is important to visualise these not as massive upwards or downwards drifts but as hinging movements involving only a few hundred feet or less of the surface here and there, leaving large areas undisturbed. The maximum amount of movement up or down was less than eight feet.

Visualising mentally the hinging nature of ground movements enables an accurate concept of what followed. As movement occurred the ground cracked. If a particular crack happened to be vertical the movement crushed it and a shaft sunk over the area today would find mixed up rock. However, if a crack happened to be oblique and the ground in the area where the crack faced upwards moved upwards, the crack would remain as an entity and appear today as a fault, or in Coober Pedy language, a 'slide'. Because of the hinging nature of ground movements, therefore, levels begin to rise away from where a slide faces upwards and reach a maximum height (about eight feet) at the slide. Levels represent areas of relative weakness, like weak mortar between bricks, so, as movement occurred some separation took place. This was widest against the slide and tapered to nothing where the levels flattened out away from the slide. At the same time vertical cracks, widest at the levels and tapering to nothing a foot or so above them occurred. These ran in a direction parallel to the slide.

Opal gel was then able to run into the spaces provided by these cracks and solidify as opal. There is seam opal in the levels and vertical opal in the vertical cracks.

If two slides cross at right angles one quadrant of the cross will be characterised by enclosing two upward facing slide faces. This is where the most ground movement occurred and the opal in that quadrant will be thicker than elsewhere. Furthermore, levels in that area will rise in two directions and verticals will not be parallel to either slide.

If multiple slides occur in an area, a small total of ground movement is shared by them all and there will be plenty of fine cracks, plenty of thin 'traces' of opal and no thick opal.

The area extending away from where a slide faces

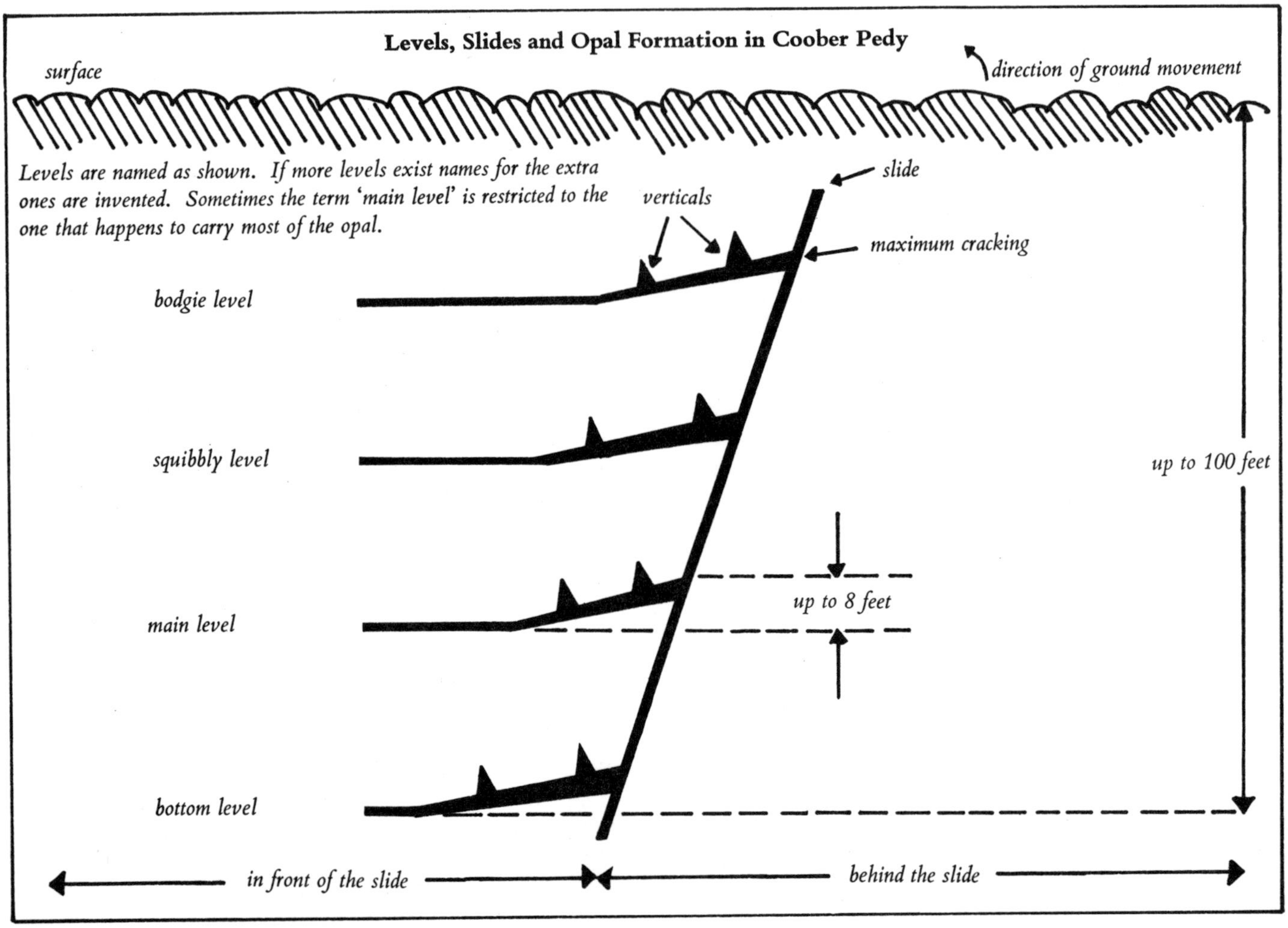

upwards is referred to as the 'front' of a slide. The other side is 'behind' the slide. Since the opal formation tends to 'run' in front of the slide for the length of a slide it is called a 'run' of opal.

Not all slides 'make' opal—most are sterile. The opal may occur in all levels or only one level; it may be limited to one end of a slide. Most of the opal formed is potch, only a little is precious opal. Some runs consist almost entirely of potch ('potch runs').

Sometimes a small, secondary slide runs more or less parallel to a major slide and cracks the ground sufficiently to permit the deposition of good opal. It is known as a 'slip' (sometimes as a 'sleeper').

The pattern of prospecting and mining in the areas under consideration can now be interpreted. First, sandstone must be found. This may begin on the surface or be covered by seventy feet or more of other rocks. Second, levels must be found. Third, rising levels are needed, and these mean slides. Fourth, drives are made in front of slides for the entire length of the slides and in all levels. If, while driving along a slide, verticals are found that do not run parallel to the slide one can assume that another slide is influencing the picture and its nature can be inferred by determining the change in direction of the verticals.

Another, rather special, feature will be discussed later and it concerns surface layers.

Knowing this geology enables one to understand the language of miners at Coober Pedy (or at least in those areas where this geology applies). There is constant talk of sandstone, of levels, of slides, of runs, and of opal.

In 1965 I knew about all this. I lived at the Six Mile, north of Coober Pedy, and often walked over the Seven Mile. In fact, I sometimes sat on it and talked the hours away. It never occurred to me to climb down some of the shafts in the area and examine the levels. I thought that the place was worked out.

But a young illiterate Greek miner (illiterate but not unintelligent) was wiser. He examined shafts in the area and deduced (by knowing what has just been discussed) that there was almost certainly a vast amount of opal in a certain place. Unfortunately for him that area was pegged and registered by someone else. So he waited, and waited, and waited . . . One day as he drove past he observed that the registered owner was sinking a shaft right over the 'guts'. His heart wept; one can imagine how he felt. Then much to his astonishment a few days later the registered owner abandoned the shaft a few feet short of the level. The waiting Greek wasted no time. He pegged the claim in his own name, finished the shaft and was rewarded by hundreds of thousands of dollars worth of opal.

The most brilliant piece of geological detective work was carried out by an older Greek, an experienced and already successful miner, in a claim that had already produced a huge fortune in opal from six levels beginning about twenty feet below the surface. The opal found in the deep levels had been thick and rich indicating considerable ground movement. The superficial twenty feet consisted of a hard, almost impenetrable, type of rock known as 'jasper'. There was opal in this layer, some had been dug out, but the effort of working it together with the danger of the inhalation of fine crystals forced the owners to abandon it (they didn't think that there was much opal there anyway).

I knew the claim well. It interested me for two reasons—the jasper that covered the sandstone, and the opal in that jasper. In fact, I photographed the mine to illustrate the red jasper dumps (compared with others that consisted of white sandstone) and I also photographed a 'specimen' of opal in jasper. Little did I realise that in my hands and before my eyes were the clues to a vast treasure of opal.

The old Greek realised it. He took the claim over after it was abandoned, reasoning that as the ground moved to split the levels in which the opal was deposited the jasper would crack open in parts too. This would not occur in levels because there were no levels. It would occur in irregular patches scattered here and there in the very hard rock. Furthermore, this suggested the manner by which it should be mined—by the rippers of a giant bulldozer.

When a large bulldozer came to Coober Pedy, the old Greek hired it. The opal came out in pockets as the rippers did their work—$20,000 the first morning, $50,000 that afternoon, opal all the way, a total weight of 320 pounds—and it could have been mine!

The operator of the bulldozer was astonished at how easy it was to find opal in Coober Pedy. As soon as his work in that claim was finished he decided to try his own luck—after all, the ground all looked the same. Needless to say, he did not find opal.

Miners at Coober Pedy know that, theoretically at any rate, all claims should be thoroughly worked, and that means all levels. Unfortunately, there is a limit as to how much work can be done. Some men tire after spending months in an unrewarding claim, others leave areas that are distinctly hopeful because another that is more hopeful turns up. Fortunes have thus been missed.

At the Old Ten Mile there was one very rich run shared by two teams. Other mines in the area were apparently sterile, including one that had been worked for nine long hard months. When the owner abandoned it, others took over, dug into the roof of a drive and recovered some of the finest opal ever seen in Coober Pedy.

At the Six Mile, in March 1965, another claim was abandoned. The previous owners had dug three 'monkey holes' in the roof without finding anything. By the end of June the claim was mine, and so was an agreeable amount of very nice opal, including a stone that weighed eleven ounces (all beautiful crystal). There were only three sterile areas in the level and they were where the previous owners had dug.

Errors are sometimes made because mines do not look promising. Levels may be ignored because they do not match accepted concepts associated with the bearing of opal.

Before leaving for a holiday in Adelaide a miner instructed men working for him to drive in a certain direction. The ground became very wet (such ground is not supposed to carry opal) and the miner admits that had he been there the claim would have been abandoned. But the workers had their orders. They kept on going—to a pocket of opal worth $60,000.

Opal may be limited in extent to one end of a slide, as one team of miners discovered for themselves. They had been following a slide for some distance. The only material produced was potch and matrix—not at all promising (they thought). When a friend arrived on the field and pegged a claim further along the direction of the slide the team did not, therefore, object. Yet, the friend's claim carried all the opal—at the very end of the slide.

The pattern of opal formation associated with slides does not apply to all areas around Coober Pedy. It has been described at length because it provides an important clue as to how opal was formed and its simplicity can be followed.

In other parts around Coober Pedy the mode of opal deposition and ground movements is not understood. At Ryan's Hill, for example, opal is found in sandstone in irregular pockets. Traces may lead into these pockets but there are no levels as such. In other areas opal stones are found mixed with surface bulldust and gravel. Some of these are of extremely fine quality—semi-black and even black. In most places opalised shells, etcetera, are common. In some areas shells occur under special rocks called 'angel stones' (different to similar named stones from other parts of Australia). Recognising these enables an experienced miner to recover shells that would be otherwise missed.

Because of the importance of sandstone in opal production it can be understood that prospectors seek it first. Unfortunately, a vast area contains sandstone and no opal. If one is fortunate enough to find opal-bearing levels breaking through to the surface, then one is lucky indeed because most of the country in the immediate vicinity of Coober Pedy has been surface prospected. However, there are vast untouched areas extending away for hundreds of miles on every side. There is no doubt that opal will be found in large amounts on new fields when prospecting is more thorough.

One can walk over a very rich field and miss it. Prospectors walked over the Eight Mile Hill near Coober Pedy for years and found nothing. Eventually an Aboriginal woman knocked over an opal-bearing stone, saw the colour and one of the richest opal fields in history was discovered. The opal mined from that small area today would be worth many millions of dollars.

There is no doubt that some Aborigines know where some of the 'undiscovered' fields are. They know the country intimately—almost every rock and blade of grass on it—but they will not reveal to white men where the opal is. They have a reason—one that we do not fully understand: it seems to be associated with tribal traditions and legends and to be very precious to the Aborigines indeed.

In spite of its apparent harshness there is something soft and beautiful about the way the Central Australian Aborigines live. The laughter of their children is an enjoyable experience. Adults commune with the country that they still regard as theirs with an understanding that white men never know. They can 'listen', in a manner of speaking, with their eyes to the sounds of nature, as if a vast and heavenly orchestra were playing. How different it is when white men dig.

The law and opal mining in Coober Pedy

When white men dig there must be laws because one cannot trust another. Laws, unfortunately, can be interpreted according to selfish individual desires, applied out of context to their original concepts, or administered incorrectly.

The purchase of a miner's right, for fifty cents, permits the owner to prospect, peg, register, and work one claim not more than 150 feet square. Work is defined as keeping 'constantly employed thereon not less than one able-bodied man in diligently searching and mining for and endeavouring to procure precious stones'. Under certain conditions a suspension of labour conditions can be applied for. If this is granted, a notice to that effect must be placed and prominently displayed on the claim. Up to four adjoining claims can be amalgamated permitting the working of only one claim at a time, provided that at least one man is employed per claim.

The holder of a miner's right can apply for a forfeit of a claim belonging to another if he considers that the claim in question is not properly pegged, registered, or worked. There are certain conditions that exempt a miner from liability to forfeiture (such as illness, attendance at a court of law, floods, or inability to obtain labour), but the onus of proof 'shall lie upon the owner of such lease'. Provisions are also made for the legal transfer of claims.

Simple though these laws may appear, they are in fact complicated. What actually happens when all the factors of geography, geology, law and human nature are mixed in the cauldron of Coober Pedy is unique in itself.

Take the ingredients and mix

A miner bottomed a shaft on good opal. There was no doubt about it, this was not just a pocket, it was a run. Elated almost beyond measure, the miner placed some charges of gelignite to commence a drive, fired them and went happily to bed while the gas cleared. Next morning he climbed down to dig out some more opal—or so he thought, because confronting him was not opal on the wall, but a vast hollow space.

It appears that the miners working next door during the previous year had discovered the opal, cleaned out the claim leaving odd pillars to support the roof and hot-footed to the safety of countries overseas. The owner's shaft had actually bottomed on a pillar. His gelignite charges had blown this away revealing the ballroom-like workings of the robbers.

When opal is found running towards another claim several things can be done. Those who discover it can inform their neighbours or keep quiet in the hope that the opal will not be discovered and the claim abandoned. Attempts may be made through forfeits, to gain legal

White opal

Jelly opal

A blaze of spectrum colours

Red covered by the blue. As this opal is moved that is what appears to happen. One of the finest opals mined

Green on blue

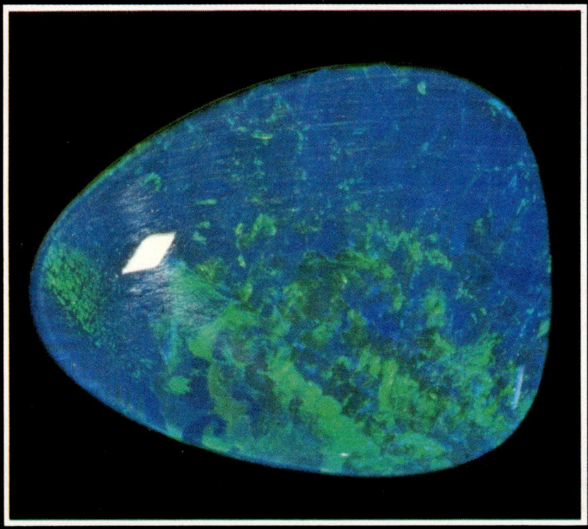

Half is brilliant, half is dull

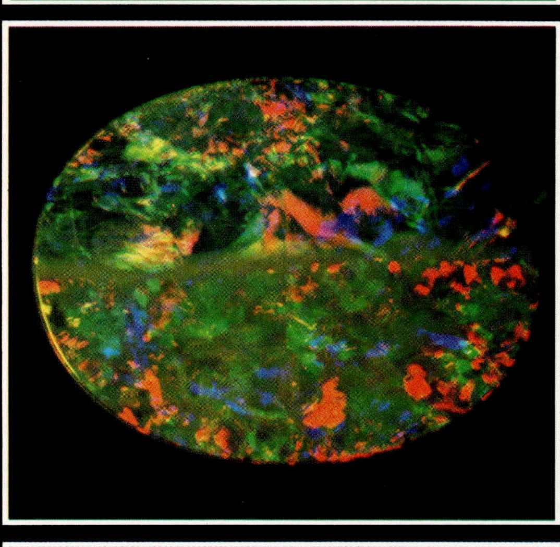

Pinfire pattern

A black opaque back and a transparent face

Crystal opal is transparent

A 72 carat crystal stone from Lightning Ridge. An exceptional specimen valued at $30,000

The red light phenomenon. A white light from one side emerges as orange or red

A slide at Lightning Ridge

Semiblack opal

White opal

Light grey opal

Dark grey opal

Black opals

Mostly crystal. An exceptional, valuable parcel

Almost semiblack—a little milky

A 'cross' between jelly and white opal

A cross' between crystal and white opal

A rumbler

Close-up of tree pinfire pattern

Almost a harlequin pattern

Hexagons rather than squares

Opal dirt at the bottom, white sandstone in the middle

Many patterns cannot be named

Solids, doublets, triplets, and chip jewellery

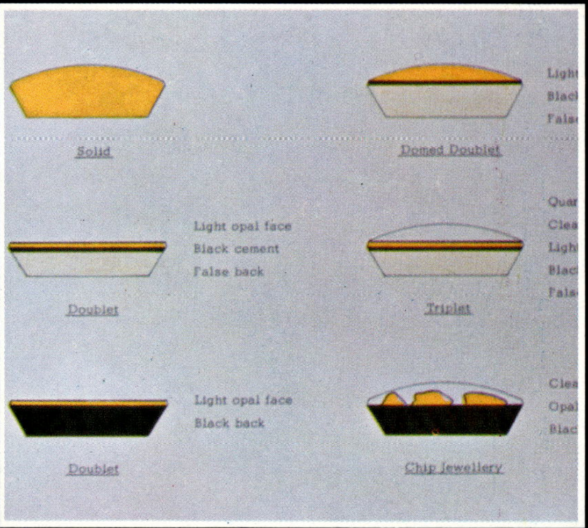

Slicing opal to make a doublet

Doublets

Chinese writing

Novelty pattern

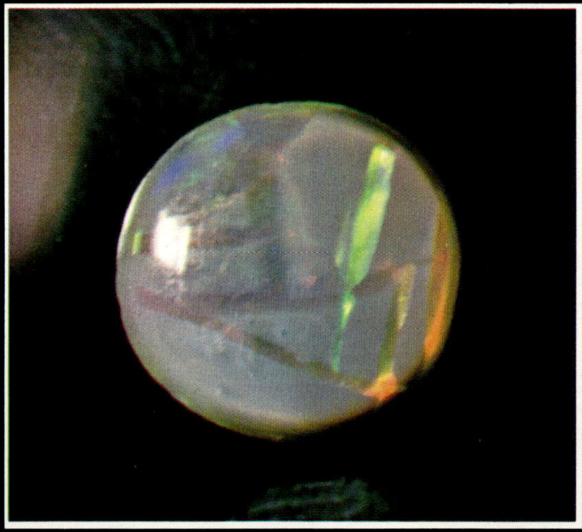

possession. This will be simple if the neighbours do not suspect the presence of opal, because they are unlikely to bother about correct pegging, registration and working. The last alternative is direct and deliberate robbery as just described.

One team of miners found opal running towards their neighbours' claim. They did the right thing and informed the men concerned. One would normally expect this to end the story but it did not. The original team became suspicious about the direction of their neighbours' drive, inquired about it and were told by the neighbours, quite frankly, that an illegal drive was being made into their claim. Rather astonished by this admission, the team requested permission to inspect. This was granted and there it was—an illegal drive, plain and clear.

What followed is not altogether clear. It is known that the team of miners collected shotguns, rifles, and sticks of gelignite, issued an ultimatum and waited. The neighbours capitulated.

Unfortunately, the outcome is not always so simple. On some of the South Australian fields gun battles have taken place in recent years, men have been murdered and others violently attacked.

Legal forfeits, at times, appear to border on the ridiculous. Some miners applied for a forfeit of a very rich claim stating that the correct number of hours had not been worked on a particular day. The onus of proof lay with the defendants and the stakes were big. Fortunately, I had helped work that claim on that day. In fact, I will never forget it, because opal nearly one foot thick came out in seemingly continuous pieces. I recorded this in my diary and used it as evidence in court. The defendants won.

On another occasion, an injunction was issued to a miner because, it was stated, his claim was not correctly worked. This injunction prohibited the owner from working his claim and entering it. Because the claim was not being worked and no notice of suspension was posted, a forfeit was asked for. Furthermore, while the court case dragged on, the contents of the claim were stolen.

A Million Dollar Game of Chess—Coober Pedy Style

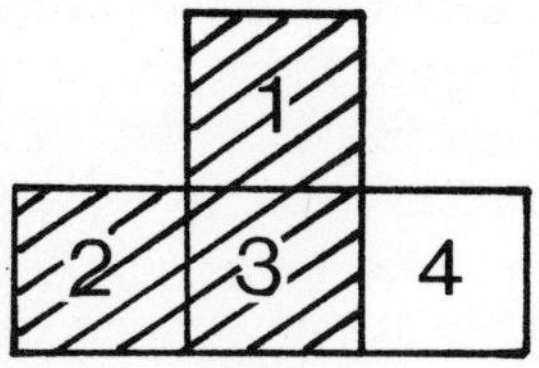

First Move: *Claims 1, 2, and 3 pegged by Anthony*
Claim 4 pegged by Nick
Anthony finds traces of opal running towards claim 4

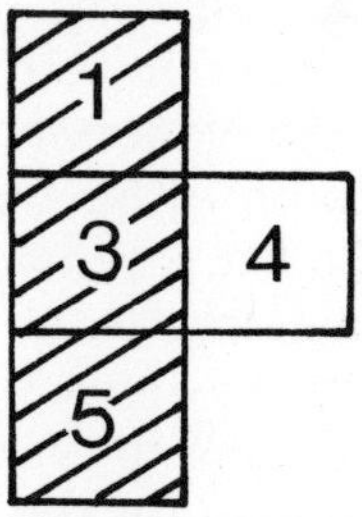

Second move: *Anthony moves his pegs from claim 2 to claim 5 thus making Nick think that he found a run passing through claims 1, 3, and 5*

Third move: *Nick moves his pegs to claim 6 in the hope of getting the tail end of the supposed run; which is exactly what Anthony expected*

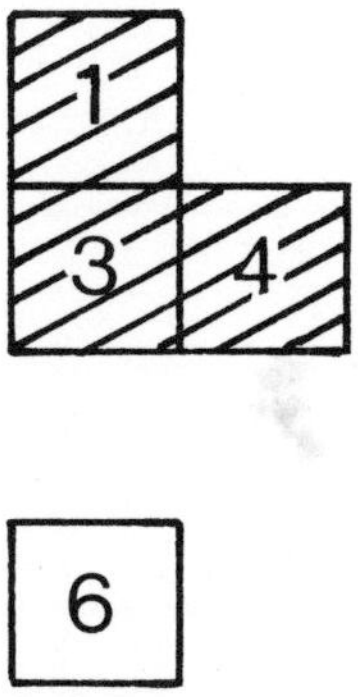

Fourth move: *Anthony pegs claim 4*

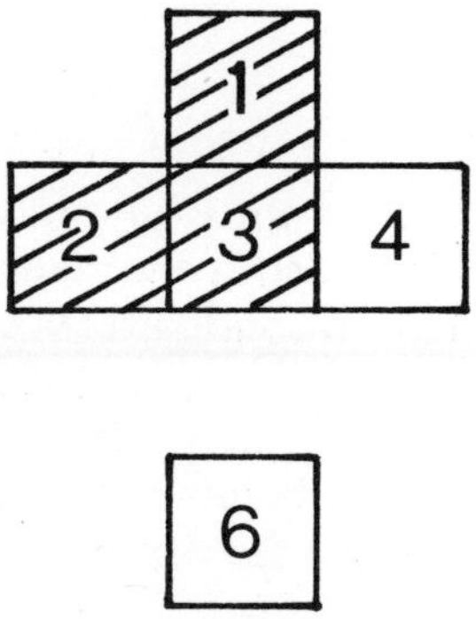

Fifth move: *'Checkmate'*
Anthony gives claim 4 to a friend
The friend gets all the opal

A million dollar game of chess—Coober Pedy style

The description that follows is basically true. Names and a few details have been changed to avoid embarrassing the men concerned.

A miner named Anthony and his two partners were working three adjoining claims pegged in the shape of a reversed 'L', the leg of which ran east and west. To the east of the south-eastern claim and directly adjoining it was one pegged by a man named Nick. Immediately to the south of this was one pegged by Nick's partner.

Anthony found traces of opal running towards Nick's claim. In order to gain possession Anthony devised a clever psychological trick. In full view of Nick he moved his pegs from the western claim and inserted them directly to the south of his other two claims. This made Nick think that a run had been found passing through Anthony's claims in a north and south direction, so he immediately pulled the pegs from his own claim and inserted them to the south of the three claims pegged by Anthony (in the hope of gaining the tail end of the imagined run). This, of course, was what Anthony expected. It was a simple matter to remove the pegs from one of his own claims and peg the one just vacated by Nick.

Nick's temper was more than a little upset when he realised how he had been tricked. But fate played the final move against Anthony. After working for a few more days, this unfortunate trickster decided that his original claim was the best one after all. He gave the claim that he had obtained so cleverly to a friend and repegged his original claim.

It so happened that the claim he gave away carried all the opal. There was opal in giant thick slabs, enough to make several men rich for life, just waiting to be dug—and Anthony gave it all away.

More about geology and psychology

Three separate miners, whose names I will change in order to protect their privacy, pegged three adjoining claims in the shape of an 'L'. The sides of the claims were roughly in an east-west and north-south direction. The northern claim belonged to Tom. It was pegged, registered, and not being worked or under suspension and therefore liable to forfeiture. Adjoining this on the southern side was Dick's claim, on the eastern side of which was Harry's claim.

Harry was a newcomer to Coober Pedy and a stranger to mining. His first shaft was commenced near to the centre of the northern end of his claim. Mystified by what he saw and not knowing what to do, he turned to me for advice. I intended to help him but something prevented this and Harry turned to another miner, named Joe.

Joe was a very astute and experienced opal miner. He climbed down Harry's shaft and saw levels rising towards the south. This meant, of course, that a slide was present only a few feet away and its direction was roughly east-west. There was also a faint possibility that the slide could carry opal.

Harry was advised to drive towards the slide. He did this and called for Joe's advice again. This time Joe recognised the slide. It was, as he forecast, running east-west. But that was not all. There were traces of opal running towards the west and minute verticals that were not running in a true east-west direction thus indicating the presence of a second slide cutting the first slide at right angles and facing towards the east.

Added up this meant, to Joe, that the opal was not in Harry's claim. It was in those owned by Tom and Dick—and there was plenty of it. Joe explained this to Harry and advised that they should form a partnership, work for a few more days, pretend that the area was sterile, move to the other end of the claim and wait.

Joe reasoned that Tom and Dick would assume that the area must be sterile because everyone knew how astute Joe was. It was extremely unlikely that anyone would think of sinking a shaft over the run because the run was right under the boundary between the claims owned by Tom and Dick. Eventually these claims would be abandoned and then, Joe suggested, he and Harry could peg them and get the opal.

Joe, of course, was relying on the fact that human psychology would prevent Tom and Dick from sinking shafts over the run. They would accept Joe's word about the sterility of the area and, anyway, who would sink a shaft on a boundary? It looks untidy!

But Harry thought that Joe thought that the opal was really in his (Harry's) claim and this was all a story to 'con' him into a share. So he did not accept the advice given.

This almost sent Joe mad. He could hear the opal, as it were, screaming at him from under the ground. To let someone else get it was too much to tolerate. Fortunately, there was still one way out, and it looked good indeed.

Tom's claim was not being worked, not under suspension and therefore liable to forfeiture. Joe raced around and surely, with trembling hands, filed an application for a forfeit. It should have been as easy as that.

But no. Joe was in such a hurry and so excited that he misread the numbers on the pegs. His application was therefore incorrect. Two days later a Cypriot arrived on the scene, heard about Joe's application and decided to apply himself 'just in case'. His application was correct.

Meantime, Harry told Dick about the slide. Dick sunk a shaft in front of it almost on the boundary and bottomed on a million dollars!

When the news of Dick's strike leaked out it shocked the field into a state of mania. 'Yesterday he was poor, today he is rich . . . It could have been ours . . . Perhaps we will be lucky too . . .' These were the thoughts of other miners, including myself, who watched with envy as Dick's financial worries ended forever.

Even more astonishing was the manner by which he was elevated in the eyes of miners to the deity. That the strike was pure luck seemed obvious but we all felt deep down that the gods had favoured Dick in a rather special way because he was a rather special sort of a guy.

Every day fuel was added to this psychological fire in massive quantities. We noticed the long hours of silence when only the hum of the generator revealed that Dick was working down below. The absence of the sound of the jack-pick and the stillness of the winch meant that opal—precious fragile material—was being dug from the level. Each hour was noted and calculated according to the opinion of the observer as $5,000, $10,000 or more.

Just as upsetting were the days when Dick did not go underground but stayed locked up in his shed cleaning opal. It was bad enough for miners like myself who had found opal; others less fortunate, who had worked for years without reward must have felt a little bitter.

When Gabriel was called to class the opal it was as if an obstetrician had been called to deliver a queen. Everyone stopped work to gaze in awe as Gabriel was admitted through Dick's hallowed door. We waited all day until he came out, blinking in the light and obviously dazed. 'Jesus Christ,' was all he said, but no more was necessary. I think that the heart of every man there missed a few beats.

Eventually, when Dick appeared in public again and spoke about his opal one fact crystalised with remarkable clarity. 'I have only the traces,' Dick declared. 'The guts is in the claim next door.' What he meant was that rich though his own mine was, the opal got thicker and better as it disappeared under the boundary. My mind began to calculate. 'If $1,000,000 is "traces" then what is the "guts" worth?'

All eyes turned to the claim in question to see who would win ownership. The Cypriot won after a long expensive battle. While this went on, there were three separate robberies of the claim's contents.

The first of these was achieved by illegally drilling a shaft with a giant machine right over the run. The men concerned were caught red-handed lifting a bucket containing opal from that shaft. The second was a story in itself.

A team of nine men drove a long underground tunnel into the claim and removed a quantity of opal. Time was short and gelignite and jack-picks had to be used. Normally this would have given the game away—other miners would have noted the direction from which the explosions came—but the robbers overcame this in a very clever way. Charges were set, ready to fire. Then all miners on the field were invited to partake of drinks in the robber's shed. While they were there, the radio or gramophone would be turned on full blast, there would be shouting and dancing, then the charges would be fired. Under such conditions it was impossible for anyone to know the direction from which the explosions came.

One day, on the Thursday before Good Friday, 1966, things nearly went wrong. After the charges were fired, one of the robbers glanced out of the window and saw smoke coming from the shaft on the disputed claim. Distances had been miscalculated and the charges had blown through to that shaft. It was necessary, therefore, to hand out a second round of drinks—and put on another record.

The men concerned in this episode were eventually apprehended and thrown into gaol.

The third robbery was planned in a different way. Two men were lowered into the mine one night by an accomplice. They intended to dig out opal and flee. But they were caught red-handed.

Mining at Coober Pedy

Puddlers and rumblers are unnecessary because the opal occurs in seams and verticals. Shallow country can be worked from the surface. Deeper ground requires the use of bulldozers, trench diggers, or shafts.

Bulldozers are economical provided that opal is there. To prospect with these can be expensive.

Overburden can be quickly removed, levels exposed and opal recovered in short spaces of time—weeks, compared with months or years using other methods. As demonstrated so effectively by the old Greek miner mentioned previously, bulldozers are the only effective means of working certain very hard types of ground, such as jasper.

Trench diggers are limited to relatively shallow country. Undermining trenches is very tempting but dangerous (it is also illegal).

Shafts may be sunk by hand (picks and shovels), with the aid of gelignite, hand drills, air compressors, drills and jack-hammers, electric drills and jack-hammers, or giant mechanical drills. A hand windlass or any form of mechanical hoist may be used to remove the dirt.

If a horizontal drive is to be made, this can be usually managed effectively with the aid of gelignite and air or electrically operated tools. Driving with picks and shovels alone is slow and tedious. Dirt may be removed in metal buckets of varying size according to the nature of hand or mechanical hoists used.

Big and efficient machinery can sink shafts quickly and remove dirt quickly. Economics depend on running costs and the element of luck.

Air supplies in shaft-worked claims can be obtained by connecting two shafts with a horizontal drive thus creating a draught. Windsocks will help or mechanical air pumps can be used.

Deep levels are usually worked first, for safety reasons and because this is easy: dirt is simply dropped down as higher levels are worked.

It is usual to remove the sandstone below levels first. Then the levels are worked. If the reverse is attempted, digging is difficult and much opal is broken or lost.

Marketing light opal

'Light,' of course, means colour, not weight. It includes jelly, crystal and some forms of grey opal.

When mined, the opal may be relatively clean but it is usually adherent to sandstone, gypsum or matrix. Removing waste material ('cleaning') is therefore the first task. When this is completed, the miner is left with up to thousands of stones varying in size, shape, and value. A classer grades these according to value (a task that may take days), leaving the opal ready for sale. Figures are quoted in dollars per ounce and vary from almost nothing to about $1,000.

Some buyers are on the field. Others are in cities throughout Australia or overseas. Many buyers visit the field from time to time.

If a miner and buyer agree, a sale may be concluded. If this eventuates, it is usually in cash. Since there are no banks in Coober Pedy, this means that enormous sums of money are carried by buyers and miners—something that in itself causes considerable trouble.

To identify parcels of opal is virtually impossible. Thefts of parcels are therefore almost as common as thefts of money, together with violence if necessary.

Rough opal may be cut on the individual fields, but usually this is not done; some opal is cut in Australia but the bulk is cut overseas. Top quality material, like good crystal and semi-black, disappears quickly because it is much in demand. In fact, it is doubtful if ordinary customers ever see it. Solids only are cut from this, doublets and triplets are made from lesser quality or poor material. Light stones may be cut by hand but the use of electronically controlled machines that cut automatically predetermined shapes and sizes is gaining favour.

When new light opal mines or fields are opened, relatively large amounts of opal are offered for sale. There is always a period of concern that lasts for several months because of the possibility of opal cracking. Once the history of the opal from a particular area is established, there is no need for concern, there is need of knowledge. Buying light opal is therefore just as difficult as buying black opal. There is no substitute for experience and knowledge.

If opal is purchased by a buyer 'in the rough', a forecast must be made as to its appearance after cutting. The opal may appear to be satisfactory, but it may crack when it is cut, or it may have minute, previously not observed, crystals of gypsum spoiling the surface. Errors in the judgment of brilliance are particularly common and financially fatal.

THE NATURE OF OPAL AND ITS FORMATION

Opal, then, consists of silicon dioxide, variable amounts of water, and traces of other elements and compounds. It resembles glass in some of its physical properties, but is fundamentally different. Compared with other gemstones it is relatively soft; fairly fragile when subjected to heat, cold, or trauma; and minutely porous. This porocity is so fine that it cannot be seen with the naked eye or ordinary microscope; it will not, in true opal, permit the injection of dyes but it will permit the loss and absorption of a quantity of water.

There are three main types of opal: *precious opal* (sometimes called 'noble opal') that exhibits a play of colour, *potch*, that does not exhibit a play of colour and is found in association with precious opal on fields that produce precious opal, and *common opal*, that does not exhibit a play of colour and is found on fields that do not produce precious opal. Since the purpose of this book is to describe the precious opal, the common opal will not be considered.

Recent work by the Australian scientists, Darragh and Sanders, has thrown considerable light on the structure of opal. Electron microscope studies have revealed that it is composed of extremely minute, sphere-like units of silicon dioxide. Between the spheres are spaces, and it seems safe to assume that at least some of the water lies free in these. If the spheres are regular in size, shape and arrangement, a three-dimensional diffraction grating is produced and this splits incident, white light into the emerging coloured rays that are responsible for the play of colour in precious opals.

If the size, shape, and arrangement of the spheres is irregular, scattering of light occurs and milky or white potch results. Other potch colours, such as black and blue, are due to traces of various elements and/or compounds, the nature of which is unknown.

Large spheres produce all the spectrum colours in the play of colour from red to violet; small spheres produce violet only; intermediate sizes produce yellow to violet.

Some emerging coloured rays, after certain critical angles of incidence to the surface (or areas of different refractive characteristics within the opal) are reflected back instead of refracted out. This is thought to be why, for example, red can sometimes change directly to spectrum blue without first changing to orange, yellow, and green.

The formation of opal

The combination of water, silica and space was essential for the formation of opal. The water dissolved the silica to form a gel which settled in the spaces provided by earth movements, and solidified as opal. Desert sandstones and opal dirt provided most of the silica, deposits of which became the opal fields, in Australia. Occasionally silica of volcanic origin was utilised, hence such forms of opal as 'mountain opal'. Space was provided in most cases by cracks in the ground resulting from earthquakes or shrinkage. Secondary spaces formed by the solution of organic materials such as wood, shells and bones, were sometimes utilised, hence opalised forms of these. Other pseudomorphs (known as 'pineapples' because of their shapes) are opal replacements of glauberite crystals.

The natural process that led to the formation of the silica gel and its solidification in the form of opal is unknown. Attempts to manufacture synthetic opal have not been successful although the sphere-like arrangement of silica particles has been copied by drying silica gel. Opal gel is found on several fields. Other evidence indicative of this being a stage in opal formation is indirect.

The mode of the inclusion of rocks in opal would be difficult to explain by anything but a gel stage. Three types of rocks can be distinguished: water-washed, rounded pebble stones; sharp-edged, rough-shaped sandstone chunks; and grains of sand or dirt. The sharp-edged stones are strongly suggestive of a rapid formation of opal; a slow formation would have dissolved the edges.

Some specimens suggest that opal gel of one type was deposited, and before this solidified, opal gel of a different type dripped into it (or was forced into it) displacing some of the original gel in characteristic styles. Some rocks in opal-bearing areas were porous enough to permit the absorption of a quantity of opal gel that set like a cement in the pores, forming the various types of true *opal matrix*.

Crystals of various types are found embedded in opal. The most common is gypsum.

The formation of patterns

Patterns in opal depend on variation in sphere sizes, regularity or otherwise of the arrangement of the spheres, block faults in the stacking arrangement, and trace materials dissolved during various stages of gel formation and solidification. It can be seen that some of these factors are responsible for patterns consisting of spectrum colours; others are responsible for potch patterns.

The mode of the formation of many patterns is not understood. However, a study of opal *in situ* has revealed some of the principles involved. They are: horizontal banding, bending of horizontal banding, drips, permeation, and concentric colour variations.

Horizontal banding is a feature of many specimens of precious opal and potch—as if the different types of opal gel tended to settle in layers. If movement of the gel occurred before solidification, the horizontal bands would be bent or distorted. Since variations in the density of potch are greater than variations in the density of precious opal, horizontal banding would be expected where different types of potch meet or where potch meets precious opal. The dripping of different types of opal gel into partially solidified opal gel has already been described. Sometimes, trace materials dissolved in one layer or portion drip or permeate into other layers or portions, giving rise to distinctive patterns. This seems to happen mainly with potch colours. Concentric colour changes are best seen in opals that are more or less spherical when mined; a common combination is a light outer shell and a dark core. How this originates is unknown.

The various methods of pattern formation are largely independent of each other: a particular specimen may demonstrate them all. Any theory advanced with regard to the formation of opal must satisfy all these features.

Some fundamental features of potch

If the 'label' of precious opal is missing, potch can be distinguished as such by the conchoidal, or shell-like, pattern on some of the fractured surfaces. Potch may be colourless, or of any colour; it may be transparent, semi-transparent, translucent, semi-translucent, or opaque. To avoid tedious repetition the word 'transparent' will be used to mean also 'semi-transparent', 'translucent', and 'semi-translucent', unless there is special reason to do otherwise.

Miners and dealers are accustomed to associating certain potch colours with those colours that form, or appear to form, the background colours in some types of precious opal. Thus, for example, white and green-white potch may look different when viewed alone but they look the same in precious opal (because most precious opals of this type have green in the colour play). It is customary, therefore, to class all near-white varieties of potch as 'white'. The same applies to other colours.

Fine differences in colour or colour combinations would be difficult or impossible to describe in practice (this includes precious opal colours as well as potch colours). Miners and dealers overcome this by a method which is, to them, very practical. They 'know' that, for example, the 'blue' from one area is different to the 'blue' from another area (mine or field), and they know the characteristics of each. Instead of becoming involved in long descriptive phrases, they simply quote the area of origin. This supplies the initiated with all the information required. The uninitiated (referred to as 'mugs', hence the term 'mug stones', or opals of such poor quality that they are only suitable for sale to those who do not understand the opal) are left to wonder. In actual fact, this intimate knowledge is not so absurd as it may seem because an understanding of faults is closely associated with it.

Following this discussion the colours in potch can be listed as: colourless-transparent, white, grey, black, blue, amber, intermediate types, and mixtures. Less common, or rare, are: red, green, orange, and others.

Colourless-transparent potch may occur alone but is usually found mixed with other forms of potch or precious opal. Compared with other forms of opal, it is light in weight and relatively fragile.

One type of potch is not colourless-transparent but slightly milky. This is due to slight irregularities in sphere size, shape, arrangement and/or traces of other materials. As the amount of milkiness increases, transparency decreases until opaque, white potch results. The same can be said for the other colours; they vary from slight tinges in otherwise colourless-transparent potch to dense, opaque shades.

Amber potch is sometimes called 'beer-bottle' potch. One variety of transparent blue (found in association with Queensland boulder opal) is called 'blue-bottle' potch. A particular white variety is called 'crockery' potch. In Queensland, large pieces of potch are referred to as 'snide'. Potch with a little precious opal mixed with it is called 'potch-and-colour'.

THE COLOURS IN PRECIOUS OPAL

A particular area on a precious opal must change colour when viewed from different angles (or when the angle of incident light alters), otherwise it would be potch and not precious opal. The colours in this colour play can be divided into two groups according to their mode of production.

1 *The spectrum colours:* red, orange, yellow, blue, green, indigo and violet. They are produced by the diffraction grating effect of the arrangement of the minute sphere-like units of silicon dioxide. Complications occur because colours can be mixed or filtered by overlying layers of precious opal or coloured potch.

2 *The potch colours:* colourless-transparent, white, grey, black, blue, amber, intermediate types, mixtures, rare types, and unusual types. In the case of colourless-transparent potch the 'colour' is produced by a critical arrangement and size of the spheres. Other colours are due to irregularities in sphere sizes and/or trace materials.

The difference between the two groups of colours is best demonstrated by what has become known as 'the searchlight phenomenon'. This was discovered (if this is the correct use of the word) by Len Cram, an opal dealer from Lightning Ridge, and myself when we attempted to demonstrate the features of opal on movie film. By adding a series of close-up lenses to a zoom lens and then zooming in close, it was found that the spectrum colours appeared as searchlights that arose from the opal and shone radially away from the edges. Potch colours did not do this.

The beauty of this demonstration is difficult to describe; coloured searchlights changing colour and moving in a most entrancing manner across the screen form a blaze of colour, or 'psychedelic'art. However, the explanation is more earthly. Coloured rays of light, emerging from the diffraction grating within the opal, enter the lens system of the camera and, because of the number of close-up lenses used, spherical aberrations occur, giving rise to the apparent searchlights. On the other hand, this demonstration does highlight the important differences between, for example, potch blues and spectrum blues, something that would be difficult to do otherwise.

Potch colours are like paints. Spectrum colours are like coloured light sources. Potch colours tend to be dull; spectrum colours tend to be brilliant.

The red light phenomenon

If a white light source is aimed at a piece of jelly, crystal or semi-black opal, the light rays emerging opposite are orange or red. This is not a play of colour but a separate phenomenon due, probably, to absorption characteristics of those particular opal types. However, the explanation remains in doubt. The richness of the orange or red light is an index of the opal's quality. In good stones some of the colour is thrown back, thus endowing them with extra beauty. If the angle of incident light is correct for a particular stone, the emerging orange rays will light up surface irregularities. Even very thin slices of opal exhibit this phenomenon; it can be seen in opals from all fields, provided they are of the types listed above.

POSSIBLE COLOUR CHANGES

Three special features will be discussed before passing on to a classification of Australian opal. These features are:

1 A spectrum colour can change to another spectrum colour (as the opal is moved) or to one, but not more than one, potch colour.

2 Some specimens never exhibit potch colours—they always show spectrum colours.

3 A potch colour can change to a spectrum colour but not to another potch colour.

The *sequence* of colour changes varies from specimen to specimen. Spectrum red, for example, may pass through every spectrum colour to violet or directly to spectrum blue. Other specimens show red changing to potch grey and then to spectrum blue. One could go on and on with many variations, all obeying the three rules listed above. This sequence influences to a very great degree the beauty of individual specimens. For example, red passing directly to spectrum blue is more attractive than red passing to potch grey and then to a spectrum blue.

The speed of colour change is also important. Red that 'holds' over widely varying viewing angles is obviously more attractive than red that 'dies' quickly into a potch colour. The first type depends partly on particular features of the opal (the arrangement of the spheres) and partly on the fact that most light sources are broad in nature. True pin-point light sources result in rapid colour changes.

'Foreground' and 'background' colours must also be considered, both in an artistic sense (for example, red tends to show up in the foreground and blue in the background) and in a three-dimensional sense as applied to opal (the red may appear to be in the background and the blue in the foreground). We have, for example, what is known as *red on blue* (spectrum red that appears to overlie potch blue), or *red on black* (spectrum red on potch black). But the full picture is even more complex. In some specimens a spectrum blue (for example) may appear to flow over a spectrum red as the opal is moved, as if a sea of blue was poured over the red. This is a particularly beautiful sequence. It is called '*red covered by the blue*'. Then there is, for example, '*red coming out of the blue*' where this actually seems to occur as the opal is moved.

No allowance has been made so far for opals that are partially transparent and therefore truly three-dimensional in colour changes.

Transparent and opaque opals

Opals are either transparent or opaque. Opaque opals can be either opaque, throughout, or they can be composed of transparent faces backed by opaque opal or rock. Obviously, all precious opals are partially transparent on the face, otherwise light could not enter and the play of colour would not be produced. However, some stones consist of obviously transparent faces backed with opaque material. These are typical examples and must be recognised as such.

The nature of the opaque back influences to a very high degree the appearance of the face. If the opal is transparent on the face and the back is light in colour, the opal looks light in colour and the colours in the play of colour tend to be weak. A similar face on a dark back looks dark in colour and the colours in the play of colour are enhanced. This is said to be due to absorption, by the dark back, of light that would otherwise be reflected back and make the colours appear weak.

There are, therefore, two types of 'black' opals. Some are black in their own right, others are black because they consist of transparent faces backed by black. The same can be said for the other colours.

THE BASIC CLASSIFICATION OF AUSTRALIAN PRECIOUS OPAL

In view of the enormous variety of the precious opal, a classification of this type may seem impossible. However, the observance of some fundamental principles enables us to give accurate descriptions and, what is more important, to establish methods of pricing. The principles are: *The dominating (or apparently dominating) potch colour determines the classification of an opal.*

The second principle is a corollary of the first, namely: *All opals that look dark, no matter how the dark appearance originates, are classed as black opals.*

These criteria enable us to build up a clear picture of the different types of precious opal. If the potch colour is colourless-transparent, and only a little colour is present in the colour play, the opal belongs to the category called 'jelly'. Usually, the first colour to appear in jelly is blue; green follows and sometimes orange and red. In most cases, however, by the time orange and red have appeared, the concentration of colour in the colour play has increased to the stage where the opal is no longer classified as jelly, but as 'crystal'. Crystal may exhibit red, orange, green and blue; orange, green and blue; or only green and blue. If the concentration of colour in the colour play is denser, yet the opal remains transparent while the blue in the colour play is moderately rich, giving the stone a semi-dark appearance, we describe it as 'semi-black'.

There is a type of opal that is more or less transparent when held up to the light and viewed from certain angles (like semi-black) but the blue in the colour play is very dark—so dark and brilliant that, when it appears, the opal is quite opaque (and extremely beautiful). This type is called 'crystal black'. In some ways it could be regarded as an advance on semi-black.

If the potch colour in an opal is light to moderately dark amber, the opal is called 'amber'. In view of our second principle of classification, dark amber opals are classed as black opals. Apart from all these, there is 'white' opal and 'grey' opal.

Opals that are opaque and contain only spectrum colours (never, from any angle of view, potch colours) will show, under careful examination, dark blue as a prominent spectrum colour. This makes these opals look dark and they are therefore, according to our second rule, classed as black opals.

The eight classes of opal are therefore: jelly, crystal, semi-black, crystal black, black, white, grey and amber.

Many stones are intermediate types or mixtures, such as the so-called 'milky' opals that are midway between white and jelly or crystal. Rare colours have not been considered.

Some jelly opals are referred to as amber. This is

Chaff pattern

Close-up of chaff pattern showing lineal striations

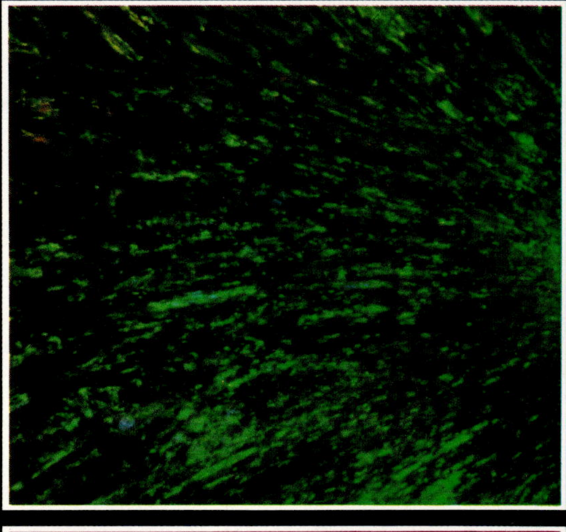

Twinkle opal

Close-up of peacock's tail pattern

Flagstone pattern

Palette pattern—like an artist's palette

Peacock's tail pattern

Broad flash

Exploding flash

A 'blow' underground at Lightning Ridge

Gouging for opal

Triplets

Chip jewellery

Seam opal showing horizontal patterns

Vertical opal showing horizontal bands

The Red Robin—17½ carats of unadulterated magnificence

The Three Mile field

The New Nine Mile field

A unique handful of nobbies from Lightning Ridge

Queensland opal matrix showing dots of opal

Matrix opal associated with seam and vertical opal from Coober Pedy (untreated)

Andamooka opal matrix before treatment

Treated Andamooka matrix

Queensland boulder opal

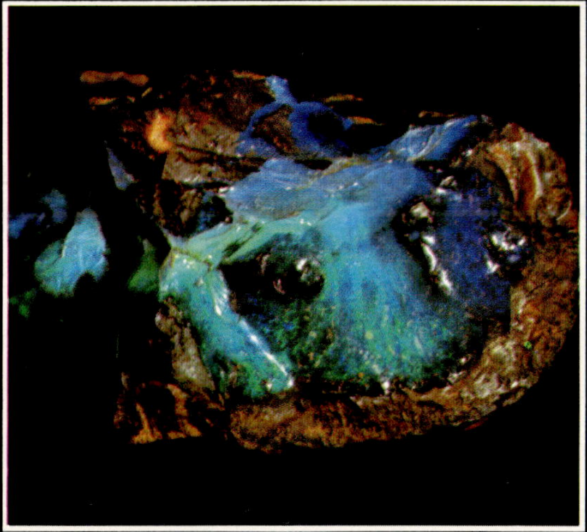

Queensland boulder opal showing 'pepper and salt' rock

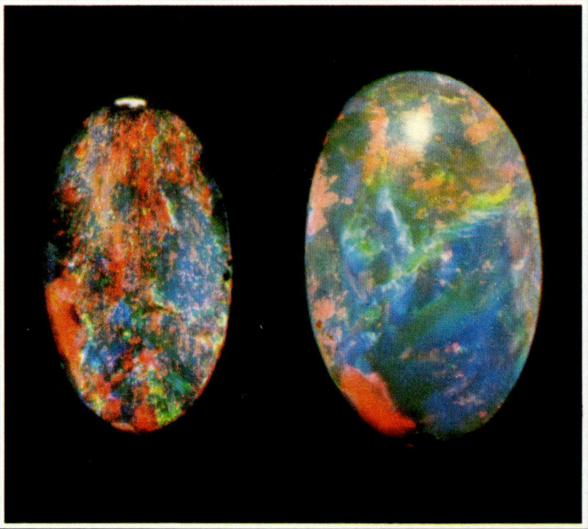

The smaller stone is Queensland boulder opal. The larger one is Lightning Ridge black opal. Note superiority of boulder red

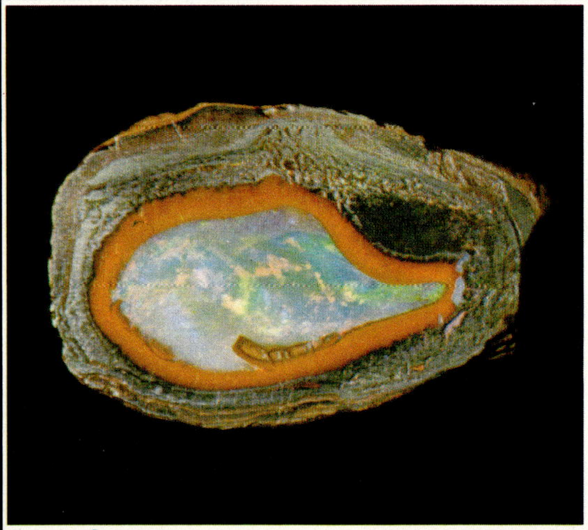

A yowah nut

Cut Queensland opal matrix stones

Cutting opal

Close-up of Queensland opal matrix

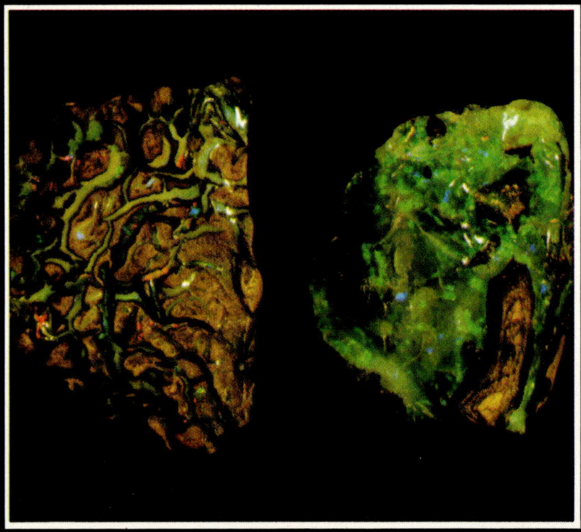

Queensland opal matrix showing sheets and veins of opal

Close-up of treated Andamooka matrix

definitely incorrect. A recent custom is the reference to so-called 'purple' stones. This term is applied to all stones that look purple—no matter how the purple colour originates—from purple-coloured potch to purple in the colour play, or even stones that are not really purple but give the impression of being so.

It would be better to class such stones basically as black and then sub-group them as 'purple' as explained in the next section. Otherwise all the principles of classification would become illogical.

Obviously, according to rule two, opals that have dark blue or dark amber in the background are classed as black opals. Various methods of sub-grouping are then used to distinguish the different types of opal or, as already described, terms such as red on blue or red on black used instead.

Sub-groups based on dominant spectrum colour in pattern

The predominance of a particular spectrum colour is a criterion for the classification of an opal in a sub-group. Many stones, of course, are mixed and cannot be specifically sub-grouped in this manner. The usual terminology is restricted to five colours: red, orange, green, blue, and purple. Other colours are usually made to fit into these five groups—violet, for example could be called 'blue' or 'purple'. However, variations are permissible, provided there is just cause. 'Green and gold', for example, is sometimes used to describe a particularly attractive crystal opal with soft green and golden-orange tinges.

'Purple' is usually restricted to rather poor quality, dark, opaque stones. As already explained, the purple may not be a true colour.

One often hears such terms as 'red stones', 'orange stones', and 'blue stones'. Such terms are incorrect. The first, for example, could be 'red on blue', 'red on black', or several others. The last could be blue jelly or black opal with a blue spectrum pattern colour, etc. It would be best to avoid the use of such terms and restrict description to correct terminology.

Brilliance in opal

A brilliant opal looks far more attractive than a dull one and is far more valuable. Degrees in brilliance are very difficult to determine and it is probable that the ability to judge this factor is important above all else.

Brilliance depends on five things: the overall intensity of the various light rays emerging from the diffraction grating within the opal; the presence of potch colours; the type of present potch colours; the proportion of potch colours to spectrum colours over different angles of view; and the amount of 'milkiness' in the opal.

The human memory is notoriously poor when degrees of brilliance are considered and because of this even recognised experts sometimes make serious errors. Direct comparisons between different stones will sometimes reveal unexpected differences.

Some opals appear to have the ability to 'concentrate' incident light in the diffraction grating and throw it out in powerful coloured rays. This is the feature that makes some opals seem to 'glow' in the dark. Of course they do not—they simply make the best use of what light is available.

Generally speaking, some types of Queensland boulder opal are the most brilliant. The 'Star of Bethlehem', a magnificent opal from Queensland, is often quoted as such an example. At Lightning Ridge, the Nine Mile field has the reputation of producing the most brilliant stones.

A STUDY OF PATTERNS

Patterns may be static or mobile, large or small. Static patterns may be likened to floor tiles that change colour when viewed from different angles, but do not move. Mobile patterns (sometimes called 'rolls') move across part or all of a stone as viewing angles alter. There is not a sharp division between the two. Many stones that apparently exhibit static patterns have individual areas that exhibit, in a miniature way, mobile patterns.

A translucent surface layer (layers are known as 'bars') may exhibit a mobile pattern that overlies a deeper static pattern; sometimes the reverse may apply. The beauty of such opals hardly needs description.

A translucent surface bar, in a domed stone, acts as a lens that magnifies underlying patterns and enhances their beauty.

Patterns tend to be formed in horizontal sheets, therefore domed stones faced in the horizontal plane (in reference to the original position in the ground) tend to have different patterns from the apex of the dome to the sides. All stones faced in the horizontal plane tend to have broader patterns than stones faced in the vertical plane. For similar reasons, stones that are translucent are more so when viewed from the side than when viewed from above.

Typical patterns are rare; atypical and mixed patterns are common; some defy description; others can only be described by enlisting the aid of special terms.

Rolling flash is a mobile flash of colour that moves across part or all of a stone as it is moved. The rate of pattern movement varies a little but it must be slow, like a wave rolling in to the shore. When the direction of the stone movement is reversed, the direction of the pattern movement must also reverse. There may be one or several rolls; they may be parallel or cross one another at right angles (very rare and beautiful); they may be narrow or broad. In black opal, a red roll followed by royal blue is the most desirable sequence. If the area following a roll goes 'dead' with a potch colour, this usually reduces beauty considerably.

There are at least two types of rolls. The first is due to the particular angle of arrangement of the spheres of silicon dioxide and is seen in opaque and translucent opals. The second is seen in translucent opals only (including triplets) and is due to the manner by which emerging light rays show on the surface.

Liquid opal is a rare mobile pattern that always moves in one direction irrespective of the direction of stone movement, like a bubble trying to escape from a liquid.

Exploding flash is rare in the large form and not necessarily confined to domed stones. The pattern explodes outwards in all directions as the stone is moved. If red is a feature and the stone is moved in the horizontal plane through 90 degrees, the red usually changes to orange or green.

Broad flash appears quickly across part or all of a stone as it is moved. Usually it consists of one colour only and is visible from only one or two angles. However, in some stones it is visible from many angles and, particularly if the flash is red, the effect of this is magnificent. The colour change can be sudden over the entire area of the pattern or it may begin at one edge and spread quickly.

Pinfire resembles a mass of pin-heads flashing different colours with movements of the stone. Sizes vary from match-head to minute.

Twinkle opal resembles stars twinkling in the sky. The pattern units are moderately small and separated to give this effect. There are grades between this and true pinfire.

Tree pinfire (sometimes called 'fern pattern') is a pinfire arranged like outward reaching branches of a tree. The 'pinheads' are often elongated along the pattern lines.

Floral pattern is so named because it resembles some of the floral patterns on printed dress materials. The units are larger than pinfire, more or less rounded but irregular. One stone from Lightning Ridge was called 'The Floral Queen'.

Flagstone pattern resembles a flagstone footpath. It is, perhaps, one of the most beautiful patterns. If the stone is small, with a central unit surrounded by a ring of outer units, it is called '*clover leaf*'.

Harlequin pattern is so named because it resembles the costume of a harlequin. The true form—squares of equal size regularly arranged—is rare. Slightly more common are units that have distorted angles, or squares that are uneven in arrangement. A radiating form, like a spider's web, is also seen. Many so-called 'harlequins' are barely recognisable as such.

Moss pattern (not to be confused with 'moss opal' which is a form of common opal) is a dull, very fine pinpoint pattern usually of a grey-blue-green colour. It is most unattractive. Needless to say, its name is derived from its resemblance to moss.

Ribbon pattern resembles coloured ribbons laid side by side. The true form is rare and best seen in Queensland boulder opal verticals. Individual 'ribbons' can change colour along their lengths either suddenly or slowly. From any particular viewing angle only one colour may be seen in each ribbon or there may be several colours; the boundaries between colour changes can be abrupt (a sharp line) or gradual (like a rainbow).

Straw pattern resembles flat pieces of straw that criss-cross each other. The amount of criss-crossing varies and the broad form is rare.

Grass pattern in some ways is like straw pattern but the units are thin like blades of grass and, typically, show lineal striations.

Chaff pattern is like pieces of chaff sprinkled onto a surface. Each unit is complete with lineal striations.

Sunflash exhibits small flashes of colour in a sea of dark potch. The colours are so weak that a strong light, such as sunlight, is necessary to reveal them.

Starflash pattern is used as a method of description if the potch body of an opal is translucent but 'inky'—that is, blackened—and the pattern units are small and relatively bright. It differs from 'sunflash' in that the latter is opaque and requires a strong light to make the play of colour obvious.

Chinese writing resembles one or more symbols (depending on the size of the stone) of Chinese writing.

Rainbow pattern is curved like portion of a rainbow. Typically it consists of pastel shades that merge slowly into each other.

Palette pattern resembles an artist's palette with dabs of paint.

Peacock's tail resembles its namesake.

Other novelty patterns include numbers, letters, scenes, portraits, eyes, poached eggs, and many others.

SOLIDS, DOUBLETS, TRIPLETS AND CHIP JEWELLERY

Solids are natural stones that have been cut and polished but otherwise not interfered with. They need not consist of solid precious opal—parts may be potch or natural rock.

Doublets are made by cementing thin slices of opal onto potch or plastic backs. Usually, for economic reasons, they are made with flat faces but domed stones are better.

If a thin slice of pale-looking opal, such as white, jelly, crystal, or grey, is cemented to a black back (or alternatively, if black cement is used—which is usually the case) the resulting doublet assumes many of the characteristics of black opal: it looks dark and the colours in the play of colour are enhanced in an almost unbelievable way. This last feature is said to be due to absorption by the black back of light that would otherwise be reflected back and make the colours appear weaker. Usually, it is possible to recognise a doublet as such because the shades of colour are characteristic.

Crystal opal makes good doublets although most cutters prefer opal with a little more 'body'. Grey opal makes dull, lifeless doublets. Occasionally, a black opal is too thin to utilise alone. It can be strengthened by being made into a doublet and, particularly if the surface is domed, its value can approach that of a solid.

Many top quality black opals are natural doublets with translucent faces and black potch backs. This does not reduce their value in any way. If anything, the reverse applies, because a good dark potch back is a good insurance against the possibility of cracking. Potch is preferable to plastic for the backs of doublets because it has the same expansion characteristics.

Triplets are doublets with domed pieces of quartz cemented onto the faces. Sometimes, with cheap triplets, plastic is used in place of quartz. The domed quartz face magnifies (and therefore enhances) patterns and protects the soft opal face.

However, the full picture is more complex and beautiful. Emerging rays of light interfere with the surface and show as reflections and rolls. This is particularly noticeable if the opal is of good quality. The quality of doublets and triplets depends on the quality of the materials used (opal, backs, and faces) and the quality of the workmanship and cement. Poor stones will fall apart or crack.

Chip jewellery utilises opal chips in various ways. A popular method is to mount them in clear plastic with black backings to enhance the colours.

Size and shape

Size is measured in two ways. Miners usually express size by weight in carats for finished stones, and ounces for rough light opal (i.e., rough, light-coloured opal from the South Australian fields). A two carat stone would make an ear-ring; a five to ten carat stone would make a ring or a brooch. Larger stones become expensive and difficult to dispose of. Manufacturing jewellers, for obvious reasons, express size by dimensions in millimetres.

The 'spread' of a stone influences, to a certain degree, its value. A stone may have a large spread (thin and flat) or a small spread (thick and domed).

Opals are cut with faces that may be flat or domed or a combination of the two (flat centres and sloping peripheries). When viewed from above they may be round, oval or any other shape. It is important that the edges be correctly shaped to accept jewellery mountings. With cheap and moderately expensive stones standard shapes and sizes are preferable because they fit standard settings. Some types of seam and vertical opal can be cut on electronic machines that automatically create standard sizes and shapes. The cost of special mountings for expensive stones is small when compared with the high cost of the stones; therefore standard shapes and sizes are not so important in this field.

Ring stones are in great demand, therefore stones suitable for this purpose are more likely to fetch high prices.

Domed stones are more attractive than flat stones for two reasons: different bars are revealed and the play of colour is enhanced by varying angles of emerging light. A flat stone needs to be of exceptional quality to catch the eye, because the viewing angles from which colour play is visible are too critical. A domed stone, by providing varying angles, is more likely to provide some colour play over at least a portion of its surface for any given angle of view.

A MINER'S VIEW OF OPAL

In the early days of mining, men who worked on various fields quickly became accustomed to different opal formations, associated rocks, and geological indications. These features, that varied from field to field, determined the amount of work involved, living conditions, the type of opal that would be found and, most important, its value.

It is not surprising, therefore, that there was a tendency to think of opal in terms of these features rather than a scientific classification. The terminology devised to suit this is still a useful method of sub-grouping (rather than primary classification). To those who are familiar with opal it conveys a wealth of immediate understanding that would otherwise require lengthy description.

Nobby opal

Nobbies are opal stones that are found only at Lightning Ridge. Sizes vary from minute to as large as a man's fist. In shape they may be rounded, oval, flattened, concave underneath with convex upper surfaces (mushrooms), concave underneath with conical upper surfaces (Chinese hats), or irregular.

Usually, they are found in more or less horizontal levels, one foot or more in thickness, consisting of the damp clay known as opal dirt. In a particular area there may be one or several levels sandwiched between other horizontal strata. Typically, the rock directly above a level consists of a more solid type of rock known locally as 'sandstone' (there are different types of rocks on various fields called sandstone and they are not necessarily similar in nature or significance), but this is not an invariable finding. Sometimes the lower portion of the sandstone is hardened into what is known as 'the steel band'. This is an indication of the presence of opal. Other indications, such as faults, vary, and their significance is not understood. Below a depth of one hundred feet, opal-bearing levels are not found.

The occurrence of nobbies in other types of rock or dirt is thought to be due to the erosion of original levels and the washing away of nobbies into their eventual resting place. The nobbies are scattered in the opal dirt (or other material) like currants in a cake but more thinly. Sometimes they congregate in groups known as 'pockets'.

It is thought that ground movements had something to do with the formation of nobbies. Another theory assumes that they are opalised marine creatures. The subject, important though it is, remains clouded in mystery.

Most nobbies are covered with a layer of opal dirt that makes their recognition difficult. To find them, a miner must chip away carefully the surrounding dirt or haul it to the surface in bulk and then sieve it in a machine known as a 'puddler'. Water is often used to make this process more efficient ('wet puddling'). The heap of gravel stones and nobbies that results is then examined for signs of 'colour'.

Nobbies may consist entirely of precious opal (rare), entirely of potch, or a mixture of the two. The precious opal may be black, but other types are common too. This is contrary to widely held views that regard all Lightning Ridge opal as black.

Crockery caps are found on some nobbies. As the name suggests, they consist of white potch Chinese hats on top of the nobbies. They are a sign of good opal if the bottom of the nobby consists of black or opaque blue potch. In such cases horizontal bands of good quality precious opal tend to occur in the body of the nobby. Many black opals cut from nobbies consist of more or less translucent faces backed with black potch. This does not in any way lessen their value. The reverse is probably true. Often, a thin layer of dark-coloured potch occurs immediately under the crockery cap. Sometimes, curtains of dark potch hang down from this and are seen in a finished opal (cut in the horizontal plane) as lines of potch in the precious opal face or as dots of potch when the 'curtains' are only threads.

Crockery bottoms are found on nobbies with or without crockery caps. Because the nobbies are white, the precious opal in them tends to be light in colour (e.g., crystal). Sand-shot faults are the curse of nobby opal. Perfect specimens are rare. Bars are common in nobbies. When cutting them it is important (and difficult) to choose the correct one as the face.

Seam opal is formed in more or less horizontal seams in levels that may occur in some types of opal-bearing rocks that vary in nature from field to field but are usually a form of desert sandstone. In a particular area there may be one level or several levels separated by six feet or more. Below a depth of one hundred feet opal-bearing levels rarely occur (except in Queensland, where depths of two hundred feet have been recorded).

Levels are common but opal is not. To 'carry' opal, the levels must have cracked open to form spaces for the deposition of the opal gel. Earthquakes and shrinkage are recognisable causes for this on some fields but not on others.

Typically, seam opal is deposited with horizontal patterns. However, distortion, including complete distortion, is common. Horizontal bars, that may be thin or over one inch in thickness, and consisting of potch or precious opal, are common. Jelly, milky opal, white opal, and grey opal, are the most common forms of seam opal (apart from potch). Crystal is rarer, semi-black is very rare, and black seam opal extremely rare except at Grawin, near Lightning Ridge. Seam opal does occur, with or near nobby opal, at Lightning Ridge.

The thickness varies from paper-thin 'traces' to slabs many inches thick. Usually, thin pieces carry the best colour and thick pieces are potch. Type and quality varies from field to field, from mine to mine, from level to level, and from portion to portion in the same level.

Included rocks are common in seam opal because of the natural tendency for pieces of rock to fall from the ceiling of cracks that were formed to permit the deposition of the opal. Seams are usually irregular so seam opal is commonly (in the rough state) also irregular in shape.

Vertical opal is usually found with seam opal, but it can also be found alone. The cracks that permitted the formation of the seams may extend vertically and permit the formation of verticals, or the cracking may occur vertically only and only vertical opal will result.

The patterns still tend to run horizontally so they run *across* the stones instead of along them. Because of the natural tendency for vertical cracks to be neat and clean vertical opal is usually neat, clean and free from included rocks. Other features are similar to seam opal.

Matrix associated with seam and vertical opal. A porous type of rock in the seams and verticals becomes impregnated with opal and matrix results. Usually, it is without colour but some specimens show beautiful colours and patterns. The matrix may occur alone but it is usually attached to potch or precious opal.

Treated Andamooka matrix (from Andamooka, South Australia). Originally this is a pale-looking porous matrix with barely noticeable opal dispersed throughout its structure. By dehydrating the stones and treating them with sugar or honey and acid, carbon can be deposited in the pores. This results in what is virtually a three-dimensional doublet. Colours are enhanced tremendously and the finished stones look similar to solid black opals.

Some of these opals are extremely beautiful. They can be distinguished by careful examination that reveals the carbon deposited in the pores. No doubt, the miner who discovered this technique must have felt like the mythical alchemist who made gold from base metals.

Queensland boulder opal. The Queensland opal fields are vast, even by Australian standards. Many of them have only been superficially worked and the possible opal content of deep strata remains to be discovered. Our information is therefore incomplete.

Queensland opal is a Cinderella. In beauty, brilliance, and sheer fascination it is unsurpassed, yet it is neglected as if it would be of less worth than its sisters from other states—which is certainly not the case.

Boulders are rocks containing variable amounts of ironstone. Shapes vary from spherical to elongated and irregular. They may be minute or over thirty feet in length and one foot in thickness. Their appearance suggests that they have been formed by a concretionary process that resulted in a (roughly speaking) onion-like arrangement of layers. Between some layers, and in cracks in the boulders, opal has been deposited.

There are two main types of boulders: *sandstone boulders* and *clay boulders* (sometimes called 'mud boulders').

On the boulder fields there is a layer of a rock known as sandstone, usually shallow, but sometimes up to one hundred feet or more deep. Below this is the clay 'opal dirt'. Sandstone boulders occur in the sandstone (usually near the bottom). They may be grey and coarse grained ('pepper and salt') or dark brown and fine-grained. The cracks holding the opal may be thick or thin. Usually, the opal in thin cracks is good quality. Opal in thick cracks tends to be potch. Vertical cracks contain opal that tends to be horizontally banded—in fact, ribbon pattern is best seen in this form of opal. Where vertical cracks turn towards horizontal cracks, some beautiful patterns result and care should be taken when cutting boulders to preserve these areas. Clay boulders are found beneath the sandstone layers in the opal dirt. They are usually finely textured and contain thin cracks that carry good quality opal.

Pipe boulder opal is found in some boulders. It may be several inches thick and a foot long. Its round section and elongated nature gives it its name. Usually, quality is very good.

Band boulder opal is found in some areas adherent to the bottom of the sandstone in sheet form. Some exceptionally fine stones have been cut from this opal. They are so named because of a hardened layer or 'band' found at the bottom of some sandstone layers.

Queensland pipe opal is found in this band. The pipes begin above it, pass through it, and hang into the opal dirt. Above the band, sand-shot faults are common; in the band, quality is best; below the band, the opal tends to turn to potch.

A special form of *seam opal* is also found in horizontal seams that sometimes occur in the band.

Queensland opal matrix is very different from other forms of matrix. The opal is distributed in irregular veins that are dispersed throughout boulders in a three-dimensional net-like fashion. The thickness of the veins varies; sometimes they are so thick that solid stones can be cut. Usually, however, it is necessary to include the ironstone in the opal. Apart from variations in the veins, the ironstone can also vary. It may be closely grained, fairly hard, and capable of accepting a high polish; or it may be 'sandy' and impossible to polish satisfactorily. There are four main types of Queensland matrix stones: *Yowah nuts, nodules, band matrix,* and *mud boulder opal matrix.*

Yowah nuts are found at Yowah, a field in the south-west portion of Queensland. They are a concretionary type of stone and may be several inches in diameter. Not all of them contain opal. The central portion may contain solid opal (in which case the quality can be supreme). The remainder of the opal is arranged in more or less concentric rings in the ironstone. Sometimes, opal 'spokes' radiate inwards from the periphery. The nuts are usually found in levels in the sandstone—particularly where dips occur.

Nodules are found in certain areas beneath the sandstone in characteristic yellowish opal dirt that runs in a more or less elongated gutter. They consist of a fine type of ironstone with fine lines of opal usually arranged in little concentrics. Typically, they hang below the sandstone like studded sausages.

Band matrix occurs with band boulder opal. Instead of forming in separate sheets it becomes mixed with the ironstone and matrix results.

Mud boulder opal matrix is found on the exterior of some mud boulders.

Finished Queensland matrix stones appear in three patterns or a mixture of two or three of these patterns: dots of precious opal in the matrix; sheets of precious opal in the matrix; irregular veins of precious opal in the matrix.

A STUDY OF FAULTS

It has been said that opals will never command really high prices because they are only 'temporary' or subject to the possibility of cracking. This is only a half-truth because the great majority of good stones will not crack unless they are treated with excessive carelessness. An expert can tell, almost always, whether or not a particular stone may crack. But the knowledge required to do this cannot be under-estimated; only a few gem dealers possess it. By revealing some of the principles involved I hope that this ability will pass on to many more.

Some faults such as sand inclusions are obvious (like smallpox scars in humans). Other so-called 'faults' such as water-bars can either spoil or enhance beauty according to circumstances (like small noses in humans).

Few stones, if any, are really perfect. The recognition of faults is important, but it is a mistake to magnify them.

Cracking and crazing

All types of opal will crack if treated harshly; some will crack more readily than others, some will crack spontaneously. 'Crazing' is a word used to describe multiple fine cracks that criss-cross in a net-like pattern. One specific factor may make one type of opal crack and another craze. For practical purposes cracking and crazing will be regarded as one.

The fundamental nature of cracking is unknown but several contributing factors could be listed as follows: drying; over-heating; and trauma.

In addition, some peculiar intrinsic factors are involved. Because of the lack of scientific knowledge, it is impossible to list these in a satisfactory manner. An attempt could be made as follows:

1 Field of origin; subdivision of field of origin; mine of origin; position in mine of origin (in that order of importance)
2 Nature of associated rocks
3 Basic type of opal
4 Nature of the back of the opal

Each of these factors must be regarded in relation to the others. For example: grey opal from one field may crack and the same type of opal from another field may remain stable.

Drying

Some types of opal crack during partial drying and it is important to realise that the degree of susceptibility to cracking varies enormously. Complete dehydration, without the destruction of the opal, is not possible. That it is not always possible to prevent cracking by soaking opal in water highlights the other factors responsible for this fault. Some dealers and collectors use glycerine in an endeavour to prevent excessive drying. There is no evidence to support or refute this practice and it is not now in common use.

Although it seems safe to assume that at least some of the water in opal lies free in the spaces between the extremely minute sphere-like units of silicon dioxide, the exact part played by it is not understood. It is therefore impossible to discuss the process of drying on a scientific basis. The subject must be approached indirectly.

Before opal is mined, it is surrounded by rocks that contain varying amounts of water and this water content is reflected, to a certain degree, in the water content of the opal. When opal is exposed to the air, it tends to lose water and this loss depends on atmospheric conditions and the type of opal. Soaking a partially dried opal in water causes it to take up water again.

No scientific criteria have been worked out until now that could help us decide whether the relative amount of water in various specimens of opal is directly related to susceptibility to cracking.

The water content of surrounding rocks is not directly related to susceptibility to cracking. Good as well as unstable opal can be found in 'wet' mines, and the same can be said for 'dry' mines. It seems obvious, however, that if the ground is particularly dry and the opal, when mined, is not cracked, it is unlikely to crack later.

The time interval between mining and cracking varies enormously. Some types of opal crack within a few days (in fact, the cracking can be heard if the opal is stored in a quiet room). Other types may not crack for many years. But the vast majority of specimens decide their own fate within a few months—a fortunate feature.

Cracking occurs, to a certain degree, before mining, as evidenced by the age-old stains seen in cracks in some specimens of freshly mined opal. Sometimes the cracking is so bad that the opal breaks up like grains of sugar (sugary opal).

Overheating

This is the result of exposure to a direct heat source or the generation of too much heat during inexpert cutting and polishing. Susceptibility varies enormously from specimen to specimen.

Reheating during recutting and repolishing must be given special consideration. Some opals seem to become more fragile with age. For this reason the task of recutting or repolishing a valuable stone should be only entrusted to an expert.

If too much heat is generated during cutting and polishing a stone may not crack immediately, but most experts agree that it may crack in the future.

Trauma

The amount of trauma that various specimens tolerate varies enormously, yet we suggest the following generalisation:

Very strong—treated Andamooka matrix; Queensland matrix.
Fairly strong—boulder opal.
Least strong—other forms.

All the other factors responsible for cracking make opal more susceptible to trauma.

It has been suggested that opal, like the metal parts of an aeroplane, is subject to fatigue: repeated minor strains accumulate and eventually only a very small knock may result in cracking. It is doubtful whether this is true but still, good opal should be protected as much as possible.

Field of origin

An intimate knowledge of the history of each field enables us to forecast, with a high degree of accuracy, if cracking is likely to occur. This is why the first question that most dealers ask when they examine a stone is, 'Where did it come from?' This is of considerable concern to opal buyers because expensive opal that cracks after it is purchased is obviously bad for business. The problem is less acute for ordinary retail customers because, by the time the opal reaches them, its stability is usually known.

Opal fields are divided into large districts, hundreds or thousands of miles apart, such as Lightning Ridge, Andamooka, and Coober Pedy. These are then subdivided into smaller areas (also called 'fields' and therefore confusing) such as the Ten Mile, Ryans Hill, New Chum, and The Phone Line. Each smaller subdivision is divided into individual mines and finally each mine is divided into 'levels' or portions.

The type of opal, the susceptibility to cracking, and the quality, vary from field to field, from subdivision to subdivision in each field, from mine to mine, and from portion to portion of the same mine. It may be mostly good or mostly bad in any of these divisions. To know opal one must be familiar with each and every one—an enormous task for which there is no substitute. To list these would be difficult and of doubtful value because new mines are constantly being opened. To purchase opal from a new mine is risky until time makes the history of the opal obvious.

Nature of associated rocks

The various rocks in which opal is formed influence the nature of the opal and its susceptibility to cracking. The reasons for this are unknown. Probably they involve trace elements and compounds, the type of water involved in the formation of the silica gel, the rate of solidification and the subsequent history of the rocks as far as ground movements and drying are concerned.

The different forms of matrix are the most stable because of the nature of the rocks forming the matrix. Boulder opal is, perhaps, the most stable form of 'free' opal. Seam and nobby opal vary to a very large extent. It is not uncommon to hear miners say, 'Good sandstone that—and good opal', or 'Bad sandstone—and the opal cracked'. What constitutes 'good' and 'bad' sandstone seems to vary. What is good in one area is bad in another.

Basic Type

First (and this must be stressed), the field of origin is important; then one could say that grey opal is more inclined to crack, good black rarely cracks, and the other types vary too much to allow generalisation.

The types of grey that crack are found on certain fields or certain mines; other types of grey are stable. Miners refer to the unstable forms as 'gray' (with a harsh accent) and the stable forms as 'grey' (with a soft accent). So deeply ingrained is this terminology that white and grey forms of poor opal are referred to as 'gray'.

Nature of the back of the opal

Rock backs as seen in some types of boulder opal are very safe because they will not crack. Good black or blue potch backs are also safe. Grey is risky unless the history of the field of origin proves otherwise. An opal that has a face tinted with a lemon colour and a grey back is dangerous indeed unless, as sometimes seen, there is a thin intermediate layer of black or blue potch between the face and the back. Solid opal that is tinted lemon is usually risky; an exception to this is boulder opal.

Translucent, or slightly translucent backs are less safe than opaque blues or blacks. Slightly translucent grey (known as 'lead') backs are the worst in this group, particularly if the lemon tinge is present.

Ginger whisker

Imagine an opal made from small cubes cemented together with fragile cement. The simple act of mining would be enough to separate some of the cubes and the trauma of cutting would be enough to separate the others. That is how I would describe this fault. In practice, the 'cubes' are not always true cubes but the angles of the pieces are always right angles.

When the stones are mined, they break apart a little but are still of reasonable size. The shape of the fractured surfaces is characteristic—sharp and right-angular where 'cubes' have been broken away. If this is recognised, the fact that further cracking, with the separation of more 'cubes', will follow attempts at cutting will also be recognised. Obviously, this fault only applies to rough opal.

Occasionally, stones that do not exhibit ginger whisker will break apart when attempts are made to cut them. In others, small pieces will 'jump' out of the surface during cutting.

Webbing

This may be apparent in a freshly mined and cut stone but usually it appears with age, beginning in one portion and 'growing' over a period of weeks, months, or years.

It is, in some ways, like a spider web, but the lines are not straight (which is one way of differentiating the fault from crazing) and tend to follow pattern lines.

It may commence on the surface, in which case it can be removed by repolishing, but it will invariably recur; or it may commence deep in a stone and spread from there.

The brilliance of each individual pattern unit is not disturbed directly, but the overall appearance is one of loss of brilliance.

Good black opals never show webbing. Most affected stones have some grey potch in their structure. Poor quality ('mug') stones, particularly the purple-blue variety, are more likely to be affected.

Scum

This is a rare fault and is not apparent in freshly cut stones. With age a film ('scum') appears over the surface, resulting in a loss of brilliance. This may be removed by repolishing, but it will certainly recur.

Potch lines

These may be mistaken for crazing or webbing lines. Fine lines of potch (sometimes referred to by miners as 'matrix' lines) are present between the pattern units, like fine lines of cement between carefully fitted flagstones.

Their origin can be recognised in uncut stones that are broken in the vertical plane. Horizontal, irregular potch bars are present. From these, fine 'curtains' of potch make their way vertically down between the pattern units. When the stones are cut in the horizontal plane (as they usually are) these 'curtains' are seen in section as 'lines'.

There is no relationship between potch lines and cracking.

Potch inclusions

If a portion of an opal's face consists of potch instead of precious opal, this will not change colour with movement; it will tend to be dull and usually its beauty is diminished. However, beauty can be enhanced in some stones when the potch inclusion adds to the 'decor'.

Sand-shot opal
Grains of 'sand' or dirt penetrate into the surface or body of an opal to produce this fault. It can weaken by making portions thin. In translucent stones it is always obvious. In some opaque stones it may not appear until cutting is carried out. A large number of otherwise good stones are spoilt by this common fault.

Gypsum-shot opal
This rare fault occurs in opal from particular portions of particular mines. Gypsum is often found near, or 'cemented' to, precious opal, but it rarely penetrates it. When it does, it is in the form of thin, elongated crystals. These will almost certainly not be recognised in the rough uncut stones. Cutting reveals them because they tend to break away, leaving minute pits in the surface.

Water-bars
These are really horizontal layers of clear potch that have been cut in the vertical plane and therefore appear as bars. In finished stones (if they appear as bars), they usually diminish beauty because they cut across the face like colourless, glassy ribbons.

However, if the stone can be cut with the face on the horizontal instead of the vertical plane, the bar will appear as a flat sheet. If this is on the surface of a domed stone it will act as a lens that magnifies and enhances patterns.

The true picture is even more complex than that. Clear potch is not clear glass and it does not have the optical properties of clear glass. It does, in fact, usually show some colour play and this, overlying or underlying other bars, enhances beauty considerably.

Cutting faults
To cut and polish opal is not a simple mechanical process because it certainly does not end with the ability to produce nice shapes and polishings. A tremendous understanding, that comes only with experience and ability, is necessary, particularly with Queensland boulder and Lightning Ridge black opal.

Cutting faults are tragic reminders of the rule that states that good opal should only be cut by good cutters.

Wrong surface or wrong bar chosen as face
A nobby from Lightning Ridge looks like a lump of dirt with flashes of colour appearing here and there. Almost certainly, part will be potch, there may be bars, and sand-shot faults will be strongly suspected. To determine which surface and bar will show the best colour, pattern and brilliance, be the most free from faults, and produce the most economical sized stone, is an art indeed.

When presented with a promising stone, a good cutter will think about the task for some time before commencing operations. He will examine it many times, make a shrewd guess as to what should be done, then lightly 'touch' it on a grinding wheel. The result will be considered carefully before the next stage is commenced. If necessary, the first guess will be modified, or scrapped, as new features are revealed. If bars are present, each one will be identified and its features noted.

Choosing the wrong surface or bar usually (unfortunately) spoils a stone and renders it unresponsive to further treatment. However, a skillful and dishonest cutter can, occasionally, deliberately make the wrong choice, producing a poor looking opal that can be purchased at a low price and later recut as a valuable gem.

Failure to recognise patterns
Sunflash is an obvious example; when the opal is cut in a poor light, it will not be seen. Some patterns show best if cut at certain angles. Others improve if the stone is cut a little more deeply.

Cutting away too much or too little
These would seem to be elementary faults but they are quite common. The first is irreparable. The second can be rectified, but only if it is recognised or the fear that a hidden fault may be revealed can be overcome.

Overheating
The friction of cutting and polishing can generate too much heat and result in cracking, immediately or later. A stone that has been cut by an inexperienced person is, therefore, always slightly suspect.

Poor polish
This is usually due to poor technique. Some stones, like Queensland opal matrix, have surfaces that contain areas of varying hardness: what will polish one portion will spoil another. This can be partially overcome by using the correct cutting materials, the correct amount of water, and the correct cutting speed.

Poor shape
Some stones cannot be cut into good shapes. Uneven surfaces and irregular shapes, however, are usually due to faulty technique. The edges must be correctly cut to accept mountings for jewellery.

Cracked opal

A black potch inclusion (does not change colour)

Crazed opal

Potch or 'matrix' lines

Gypsum-shot opal

Sand-shot opal

Webbing

Wet puddling

Driving beneath a sandstone roof in one of the richest claims in Lightning Ridge history

Lightning Ridge and the New Town field

Dry puddling

Fine puddled dirt compared with unpuddled material

A giant drilling rig

A power hoist

An open cut

Timber props

A slide underground

A slide breaking through to the surface. Note different rock colours

A rising level—the darker material near the 'roof'

Opal in jasper—from the mine discussed in the text

The author's mine at The Six Mile

A pure crystal stone from the author's mine. Weight eleven ounces. Value at the time $5,000

Opal may be limited to one end of a slide. This is the claim mentioned in the text

The site of the million dollar game of chess

The mine mentioned in the text

A dealer inspecting Coober Pedy opal

A bogger at work underground at Coober Pedy

Mine Tunnel Enquiry

Woomera police are investigating an alleged illegal mining operation at Coober Pedy which culminated in an explosion in a mining shaft on April 16.

It is alleged a long tunnel was dug 60 ft. down and 45 ft. into an adjoining claim, the ownership of which is under dispute before the SA Supreme Court.

The tunnel is understood to have penetrated 45 ft. into the disputed claim along two legs which diverged from a central tunnel at the claim boundary.

Police believe the tunnel was exploded in an attempt to destroy incriminating evidence.

STOP PRESS

The newspaper cutting mentioning the robbery

HOW TO VALUE OPAL

That an opal is only worth what a customer is prepared to pay, is a business principle that does not consider what it *should* be worth. Ignorance, fears of faults, and superstition, contribute to the peculiar fact that nature's most beautiful gemstone is not man's most precious.

To the uninitiated, pricing individual stones is an impossible task. There are so many various features and so much knowledge is necessary that only experts can accurately determine values. Even then, there may be considerable variations in opinions. But it is a fact that most experts agree very closely; the more expert they are, the closer their agreement will be because of their understanding of the opal's features and of the available markets. A computer-like mind is necessary to store details of stones and recall them when examining new ones.

It is also true to say that, in some dealer partnerships, business can be conducted by letter or phone—a few words of description is all that is necessary to convey the mass of detail required. By summarising the principles involved I will try to demonstrate how this is done.

To avoid complexities, each feature will be considered alone; but it must be remembered that the final value is a computerised version of all the features plus a consideration of the available markets.

Only a summary will be given because to do otherwise would mean repeating in detail what has already been discussed.

1 First, consideration must be given to the mode of presentation (in the following order): domed solids, flat solids, domed doublets, triplets, and flat doublets.

2 Next, the basic type: black, crystal black, semi-black, crystal, white, grey, and jelly (following this order). Amber must be valued separately while the dark varieties such as black opal and others will be valued according to appearance.

3 Treated Andamooka stones and Queensland opal matrix stones are valued according to appearance.

4 Brilliance is important. The absence of 'milkiness' adds value. It is important to know how some pattern areas 'glow' and 'throw out' strong rays of light. The difficulties in determining the brilliance cannot be overstressed. When contemplating the purchase of expensive stones, comparison with 'standard' stones and a 'thinking' period (delay and reconsideration) are wise precautions.

5 Big patterns are preferable to small ones. Rare, pure, and interesting patterns have increased values.

6 Red in the colour play is preferable to other colours which are, in order of value—orange, green, and blue.

7 In black opal, red followed immediately by royal

blue (as the opal is moved) is preferable to other sequences of colour change.
8 The 'trueness' must be determined. A 'true' stone does not go 'dead' or show potch colours from some angles of view. The less potch colours are apparent, the more valuable the stone.
9 The ability to remain beautiful in various types of light (direct sunlight, open shade, deep shade, and artificial light) influence the value for obvious reasons. There are 'night' stones, 'sun' stones, and 'shade' stones where the best features are revealed in the particular type of light.
10 Size and shape must be considered. If a stone is too small (less than one carat) it cannot be used alone. If it is too large (over ten carats) potential customers become too few. Standard shapes and sizes are preferable (except when expensive stones are considered, because in such cases the cost of special mountings does not matter) and so are 'ring' stones (because of demand).
11 The absence of faults increases value. The back must be good and, if possible, the history of the field of origin ascertained so as to reduce the fear of future cracking.
12 Rough opal should be considered separately. Uncut nobbies from Lightning Ridge are rarely sold because of the difficulties in determining in advance the condition of finished stones (especially the presence or absence of sand-shot faults). Some miners cut their own nobbies while others employ professional cutters who usually provide expert opinions as to the value of stones.

Seam opal from Andamooka and Coober Pedy is mostly sold in the rough state by the miners to professional buyers. Before this is done, the opal is 'classed' (graded and valued) by various expert 'classers'. There may be hundreds or thousands of rough stones in what is known as a 'parcel'. The classer sorts these into heaps according to value. Any stone that needs rock or potch snipped away is dealt with during this process that may take several days. Values are based on quality, size (which influences the amount of waste during cutting), a knowledge of the field of origin (freedom from the possibility of cracking and faults), a forecast of the appearance of finished stones, and a knowledge of correct values. At the time of writing, prices range from a few dollars per ounce to one thousand (rarely more).

Rough Queensland boulder opal is difficult to deal with because thick specimens of precious opal are rare and much depends on the ability of the cutter who can, if he is an expert, cut extremely beautiful and valuable stones from thin irregular seams embedded in thick hard rock.

It must be stressed here that in the final consideration some doublets (for example) will be worth more than some solids, and some grey opals worth more than some black opals.

What does an expert do when he considers the purchase of an opal? First, he tries not to use his imagination (because opals are like pretty girls: one can even fall in love with a particular stone and be quite blind, temporarily, to the most obvious faults). He will try to judge the stone by what he can see, not by what he would like to see; then he will quickly examine the face, carefully check the back and turn once more to the face. Usually he will deduce the field of origin; if not, he will ask about this, and also about the weight. Into his vast store of knowledge and experience he will feed the details and produce an answer in dollars per carat. This will vary from less than one to more than eight hundred. Stones in the extreme upper bracket are usually referred to as 'gems'.

Crystal gazing into the future of opal

Ten years ago there was very little classified knowledge concerning opal. Now, as readers of this book will realise, the situation is vastly different. New knowledge and a better understanding of old knowledge has removed once and for all the mysticism surrounding the world of opal. We have a proper basis for description, classification and evaluation. We know how the colour play originates, we understand the three dimensional arrangement of colours, we know about the mobility of patterns and we have come to grips with the principles of geology. What remains?

The most tantalising problem is the production of cultured, or synthetic precious opal; that is, as nature has made it. The process cannot be too involved. We know that silica gel is necessary and that this must be dried. It is possible that traces of other elements and/or compounds are necessary, but if this is so their nature remains mysterious. In fact, we do not even know why potch is sometimes black. This apparently simple problem defies explanation.

As far as the Australian opal industry is concerned, the need is not so much for new knowledge but for the spread of what is already available. Top quality stones from Lightning Ridge are only mined at a rate of two or three each year. Miners and some dealers know what their values should be, but this is of no use in a world that does not recognise true uniqueness and beauty. In this regard, a register of opal features should be kept. How many stones, for example, exhibit double rolls with one roll crossing the other at right angles? Nobody knows. Unless the buying public have some method of checking a dealer's opinion, prices will remain far below what they should be.

I wonder what prices would be if everyone saw what I do when I look into an opal. I see not just flashing colours and patterns but men and a country combined in splendours. I see the desert and its legends, the nearness of the stars, the glory of the ranges and the greatness of the plains. I see the lives of many and I see the way they died. It's all a part of nature and if others know this, they will see it too. Who cares about the price of opal? Just let me hold it in my hand.

INDEX

Aborigines 11, 13, 19, 22
Alluvial opal 14, 16
Amalgamation of claims 22
Amber opal 29, 34
Amber potch 29
Andamooka 14
Andamooka matrix 41
Angel stone 21

Background colours 32
Back of opal 33, 45
Band boulder opal 42
Band matrix 42
Bars, in opal 36, 41
Beer-bottle potch 29
Behind slide 11, 20
Blue-bottle potch 29
Biscuit band 11
Black opal 8, 9, 11, 33, 34, 41
Blow 11
Bodell, Fred 11, 12
Boulder opal 42
Brilliance 16, 26, 35
Broad flash 37
Bulldozers 16, 21, 25
Butterfly 10
Buying opal 16, 26, 44

Cardinal 9
Chaff pattern 37
Chinese hats 40
Chinese writing pattern 37
Chip jewellery 39
Claims 15, 22
Classing opal 26
Clay boulders 42
Cleaning opal 26
Colour changes in opal 32
Colours in precious opal 30
Common opal 27
Coober Pedy 8, 14, 18
Cracking of opal 26, 43
Cram, Len 30
Crazing of opal 43
Crockery bottoms 41
Crockery caps 41
Crockery potch 29
Crystal black opal 34
Crystal opal 26, 34
Cutting faults 46
Cutting opal 26, 44, 46

Dealers 16, 17
Dome in sandstone roof 11
Domed doublets 38
Dominant spectrum colours 35
Doublets 38
Drives 20
Drying of opal 43

Eight Mile field 22
Electron microscope studies 27
Exploding flash pattern 36

Faults, geological 10, 19, 40
Faults in opal 43
Fern pattern 37
Fields 8, 12, 15, 40, 44
Flagstone pattern 37
Floral pattern 37
Floral Queen 37
Foreground colours 32
Forfeiture of claims 22, 23, 24
Formation of opal 21, 27, 28
Front of slide 11, 20

Geology 10, 12, 19, 24
Ginger whisker 45
Gouging 12, 14
Grass pattern 37
Grawin field 41
Gray opal 45
Grey opal 26, 34, 45
Ground movements 10
Gypsum 26, 28, 46

Harlequin pattern 37
Heating of opal 44, 46
History of Lightning Ridge 11
Hodge, Harold 6, 12
Hoist, automatic self-tipping 14, 16

Idriess, Ion 12

Jasper 21
Jelly opal 26, 34

Kangaroo court 12
King slide 10

Law and opal mining 15, 22
Lead backs 45
Level 10, 16, 19, 20, 21, 22, 24, 26, 40, 41
Lightning Ridge 8, 9, 18, 35
Light of the World 10
Light opal 8, 14, 26
Liquid opal 36

Matrix 11, 21, 28, 41
Milky opal 34
Miner's right 15, 22
Mobile patterns 36
Monkey hole 21
Moss pattern 37
Mountain opal 28
Mud boulders 42; opal matrix 42
Mugs and mugstones 29, 45
Mushrooms 40

Named stones 9
Nature of opal 27
Nettleton, Charlie 11
Nine Mile field 35
Nobbies 12, 16, 40
Noble opal 27
Nodules 42
Noodling 15

Opal dirt 10, 40
Opal gel 8
Opal matrix 28
Opaque opal 33

Palette pattern 37
Pandora 10, 12
Parcels of opal 26
Pattern formation 28
Patterns 25, 36
Pay dirt 12, 14
Peacock's tail pattern 37
Pegging claims 15, 22, 24
Pillars 12
Pineapples 28
Pinfire pattern 37
Pipe boulder opal 42
Pockets 12, 21, 40
Porosity of opal 27
Potch 20, 21, 27, 28, 29
Potch-and-colour 29
Potch colours 29, 30
Potch inclusions 45
Potch lines 45
Precious opal 20, 27
Preston's stone 9
Pride of Lightning Ridge 10
Prospecting 14, 15, 20
Psychology of mining 24
Puddling 14, 16, 25, 41
Purple stones 35

Queensland 8, 41
Queensland boulder opal 42
Queensland opal matrix 42
Queensland pipe boulder opal 42

Rainbow pattern 37
Ratters 12
Red coming out of blue 32
Red covered by blue 32
Red light phenomenon 31
Red on black 32
Red on blue 32
Red Robin 9, 15
Registering claims 15, 22
Ribbon pattern 37, 42
Robbery 23, 25
Rolling flash 37
Rolls 9, 36
Rough opal 26
Rumblers 16, 25
Run, of opal 8, 18, 20, 24
Ryan's Hill 21

Sand-shot opal 16, 41, 43, 46
Sandstone 10, 18, 19, 20, 21, 40
Sandstone boulder opal 42
Sandstone roof 10, 11, 12, 16
Scum 45
Seam opal 19, 41
Searchlight phenomenon 30
Selling opal 16, 17, 26
Semi-black opal 34
Sequence of colour changes 32
Shafts 12, 18, 26
Shells 21
Shincracker 10
Silica 10, 20, 28
Size and shape 39, 46
Sleeper 20
Slide 10, 11, 19, 20, 24
Slip 20
Snide 29
Solids 38
Space for opal formation 28
Spectrum colours 30, 35
Speed of colour changes 32
Spiders 12
Star of Bethlehem 35
Starflash pattern 37
Static patterns 36
Steel band 10, 40
Stowed dirt 12, 14
Straw pattern 37
Structure of opal 27
Stuart Range 18
Sub-groups of opal 35, 40
Sugary opal 44
Sunflash pattern 37, 46
Surface dirt 14
Suspension of labour conditions 22, 23
Synthetic opal 28, 48

Tanks 12
Terminology 7, 9, 11, 40
Toe dirt 12, 14, 16
Traces 12, 19, 24, 25, 41
Transparent opals 33
Trauma, effect of 44
Treated Andamooka matrix 41
Tree of Knowledge 13
Tree pinfire pattern 37
Trench digging 16, 26
Triplets 38
Trueness of opal 48
Twinkle opal 37

Valuing opal 47
Vertical opal 41
Verticals 19, 20, 24

Wall 11
Wallangulla 11
Water bars 43, 46
Webbing 45
White Cliffs 8, 11
White opal 34
Windlass 12
Windsock 26
Working claims 15, 21, 22

Yowah nuts 42